Sound Recording and Reproduction

BASIC BETACAM CAMERAWORK
Peter Ward

BASIC FILM TECHNIQUE
Ken Daley

BASIC TV REPORTING
Ivor Yorke

BASIC TV TECHNOLOGY
Robert Hartwig

CONTINUITY IN FILM & VIDEO
Avril Rowlands

CREATING SPECIAL EFFECTS FOR TV & VIDEO
Bernard Wilkie

EFFECTIVE TV PRODUCTION
Gerald Millerson

GRAMMAR OF THE EDIT
Roy Thompson

AN INTRODUCTION TO ENG
Bernard Hesketh
Ivor Yorke

LIGHTING FOR VIDEO
Gerald Millerson

LOCAL RADIO JOURNALISM
Paul Chantler & Sim Harris

MOTION PICTURE CAMERA & LIGHTING EQUIPMENT
David W. Samuelson

MOTION PICTURE CAMERA DATA
David W. Samuelson

MOTION PICTURE CAMERA TECHNIQUES
David W. Samuelson

THE PANAFLEX USER'S MANUAL
David W. Samuelson

THE PHOTOGRAPHIC LENS
Sidney F. Ray

SINGLE CAMERA VIDEO PRODUCTION
Robert Musburger

16 mm FILM CUTTING
John Burder

SOUND RECORDING AND REPRODUCTION
Glyn Alkin

SOUND TECHNIQUES FOR VIDEO & TV
Glyn Alkin

THE USE OF MICROPHONES
Alec Nisbett

VIDEO CAMERA TECHNIQUES
Gerald Millerson

THE VIDEO STUDIO
Alan Bermingham, Michael Talbot-Smith, Ken Angold-Stephens & Ed Boyce

VIDEOWALLS
Robert Simpson

YOUR FILM & THE LAB
L. Bernard Happé

Sound Recording and Reproduction

Third Edition

Glyn Alkin

FOCAL PRESS

Focal Press
An imprint of Butterworth-Heinemann
Linacre House, Jordan Hill, Oxford OX2 8DP
A division of Reed Educational and Professional Publishing Ltd

A member of the Reed Elsevier plc group

OXFORD BOSTON JOHANNESBURG
MELBOURNE NEW DELHI SINGAPORE

First published 1981
Reprinted 1987 for the first time in the paperback Media Manual series
Second edition 1991
Third edition 1996

British Library Cataloguing in Publication Data
A catalogue record for this book is available from the British Library

ISBN 0 240 51467 X

Library of Congress Cataloguing in Publication Data
A catalogue record for this book is available from the Library of Congress

Composition by Genesis Typesetting, Rochester, Kent
Printed and bound in Great Britain by
Biddles Ltd, Guildford and King's Lynn

Contents

Copyright considerations

It should be emphasized that the recording of broadcast material, especially if it involves music or scripted speech, is almost certainly an infringement of copyright, both as regards the artists or recording company concerned and the broadcasting authority.

Broadcasting organizations tend to turn a 'blind eye' to domestic recording for personal use but the reproduction of copyright material for sale or public performance is a serious offence unless the permission of the holder of the copyright or their agent (in UK usually the Mechanical Copyright Protection Society) has been obtained and a royalty paid (typically 6¼% for each copy).

Introduction

The book is intended to bridge the gap between the professional recording engineer and the enthusiastic amateur, who wishes to improve his technique and gain a better understanding of the recording medium and sound quality in general.

The process of producing and recording sound is a fascinating amalgam of art and science. It is not just a matter of reproducing the original sound as accurately as possible, but of tailoring the reproduction to suit the recording medium and the likely environment in which it is to be heard. In general, the technique can be divided into two categories:

1. The recording/reproducing function, where the aim is for the recording equipment to be totally transparent, i.e. the signal should be recorded and reproduced with absolute fidelity, nothing being added, altered or taken away.
2. The sound production function: adapting the sound for the best artistic effect, taking into account the requirements of the recording medium and listening conditions as well as such matters as style, listener preference and fashion.

Even in those circumstances where it might be assumed that the ideal would be to reproduce exactly the sound in the concert hall or studio, the result would be unlikely to suit the listener's domestic listening conditions, bearing in mind the need to restrict the volume range for a small room and the effect this would have on the balance and character of the sound. Co-operation between artist and engineer can produce new and exciting art forms that improve on reality.

It usually comes as a surprise to people, when starting to use microphones, to discover how different the reproduced sound can be from the original, even when the equipment is impeccable. To understand the reason for this it is necessary to have some knowledge of the nature of sound, the effect of acoustics, the process of hearing and the mechanism of microphones.

The book follows the course of the production of a recording, from the acoustic environment and production techniques through to the various methods of sound recording and reproduction, dealing with each subject in a practical manner and concentrating on those aspects that are of particular significance to people engaged in the art and science of audio production and recording.

Sound recording is a highly complex business, particularly in relation to digital techniques, so the book includes some theory (where it is desirable for the better understanding of the medium) in as practical and non-mathematical a manner as possible.

Sound is an aural sensation created by physical vibration.

The Nature of Sound

Before starting to consider the process of sound recording, it is worthwhile pausing briefly to refresh our minds about the nature of the medium that we wish to capture.

What is sound?

Sound is an aural sensation stimulated by a vibration or mechanical wave motion in matter, operating within the frequency range that we can hear.

The simplest forms of sound vibrations, into which all sounds no matter how complex can be resolved, are pure tones such as can be produced by tuning forks. Tuning forks produce simple harmonic motions, i.e. motions which could be described by plotting the displacement of a swinging pendulum on a graph against time. The resulting graph is the familiar sine wave curve in which a complete cycle of vibration is indicated by an excursion both above and below the base line and return to the equivalent position.

How sound is propagated

Wave motion can be *longitudinal* (when the direction of motion is the same as that of propagation), *torsional* (a twisting motion) or *transverse* (as with the vibration of strings).

Sound waves require a material medium for propagation, e.g. air, water or wood. Unlike light, sound cannot travel through a vacuum.

When sound travels through the air, which is the way it usually reaches our ears (it is also possible to hear through direct mechanical contact by bone conduction), the wave motion is longitudinal. The waves consist of variations in air pressure, oscillating alternately above and below the prevailing barometric pressure. They travel outwards from the source at a speed of about 340 metres per second.

Velocity of sound

The actual velocity of sound waves depends upon the atmospheric pressure and the density of the air through which they pass. This is given by the equation:

$$v = \frac{1.41\,P}{D}$$

where v is velocity in cm/sec, D is the density of the medium in g/cm^3 and P is the barometric pressure in dyne/cm^2.

The velocity of sound increases with heat (which expands the air thereby reducing its density) and with humidity (since water vapour has a lower density than dry air).

Sound propagation through solids tends to be much more rapid and efficient than through air; for example, through steel it can travel about fifteen times as fast.

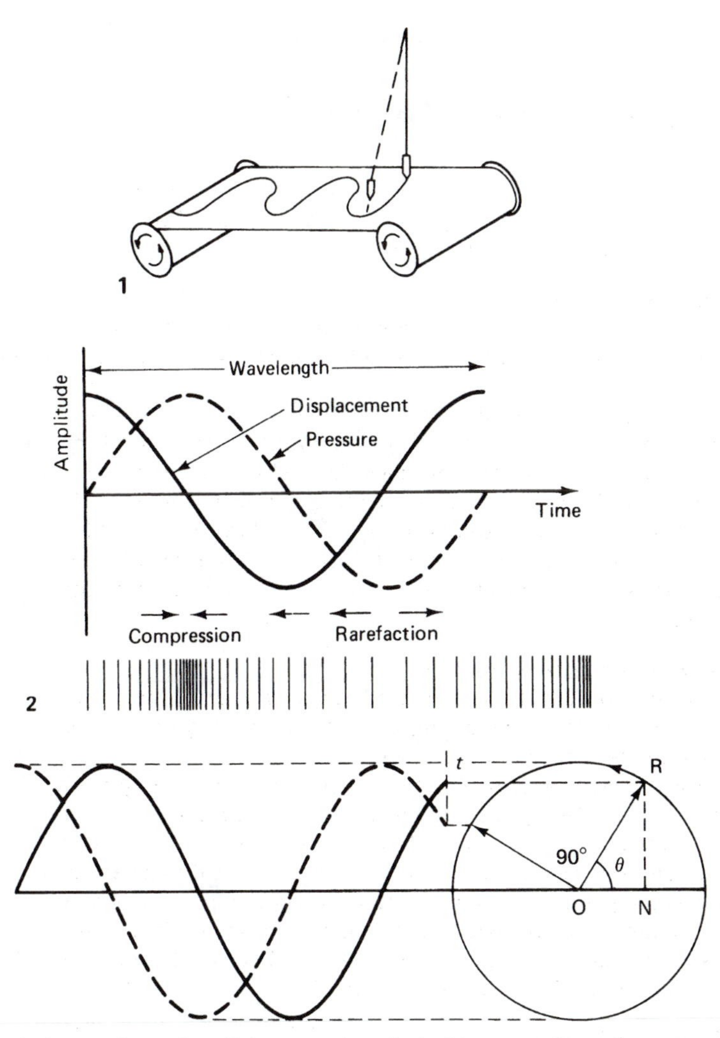

1. Simple harmonic motion. If the paper is pulled with a smooth motion past a pendulum swinging in one plane, the trace will be a simple harmonic (sine wave curve). The number of times the curve repeats itself per second is the *frequency*.

2. Sound is transmitted through the air by a series of alternating compressions and rarefactions, i.e. changes of air pressure above and below the prevailing atmospheric pressure. Maximum pressure variations occur during maximum *change* in displacement of the exciting force, i.e. when crossing its normal state. Pressure variations are minimum when the force is at a maximum.

3. Wave motion can be represented by rotating vectors. The radius OR represents the amplitude of the wave and the resultant value above or below the base line is RN. As the vector rotates plotting the angle θ against RN is equivalent to plotting θ against $\sin\theta$, hence the curve is called a sine wave. The diagram illustrates two sine waves with a relative phase angle of 90° at time *t*.

Sound waves are described in terms of amplitude, frequency and wavelength.

Sound Characteristics

Soundwaves can be defined as those with a frequency range that lies within the audio spectrum.

Frequency

The rapidity with which a cycle of vibration repeats itself is called the frequency. The audio frequency range is generally considered to extend between about 15 Hz and 20 kHz (one hertz (Hz) = one cycle per second). In practice the actual range of sounds that we can hear varies considerably between individuals, particularly at the high frequency end of the scale, where aural sensitivity tends to diminish with age after about 25 years.

Wavelength

Having established that sound travels with a certain velocity and that each point of repetition of the waveform passes a given point with a certain frequency, it follows that these points must be a certain distance apart. This distance is known as the wavelength (λ). It is related to frequency (f) and velocity (v) by the simple equation:

$$\lambda = \frac{v}{f}$$

Many of the major problems in sound engineering stem from the wide variations in wavelength that comprise the audio spectrum, i.e. wavelengths ranging from under 2 cm to about 17 m.

Amplitude

The strength of a sound wave is known as its amplitude. This is a measure of the extent that the wave, i.e. the pressure change in air, deviates from the normal state at its maximum value.

Loudness

When sound reaches the ear, the loudness we hear is not strictly proportional to the energy of the sound wave. The sensitivity of our hearing depends upon the frequency and the intensity of the sound.

Just as our appreciation of musical pitch is determined by the ratio rather than the numerical value of the frequency interval, e.g. the frequencies at either end of an octave are always in the ratio of 2:1 regardless of actual numerical difference, so the difference in loudness depends upon the ratio, rather than the numerical difference in intensities. For this reason relative intensities are expressed in bels (B). This is a unit that compares the logarithm of their powers to the base 10. In practice the bel is too large a unit and the decibel (dB) is used. This represents about the smallest discernible change in audio power, equivalent to about 26%.

$$\text{decibels} = 10 \log_{10} \frac{P_2}{P_1}$$

See Appendix 1: The Dangers of High Sound Levels.

Sine wave curve representing the variations of pressure with distance for a pure tone.

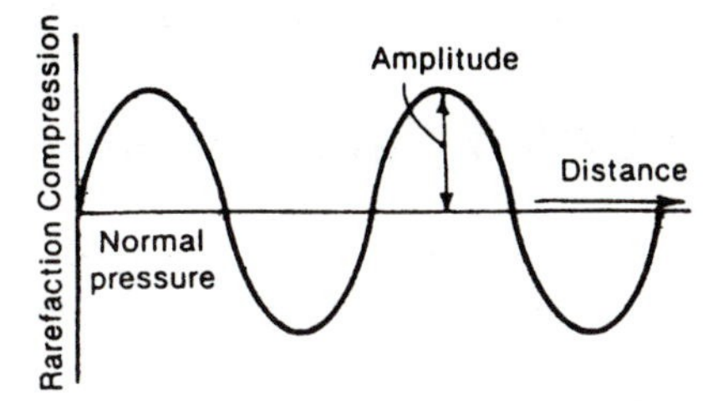

Sound waves have a large range of physical dimensions comparable to many items of furniture and equipment etc. Sound is capable of bending around obstacles whose dimensions are smaller than the wavelength.

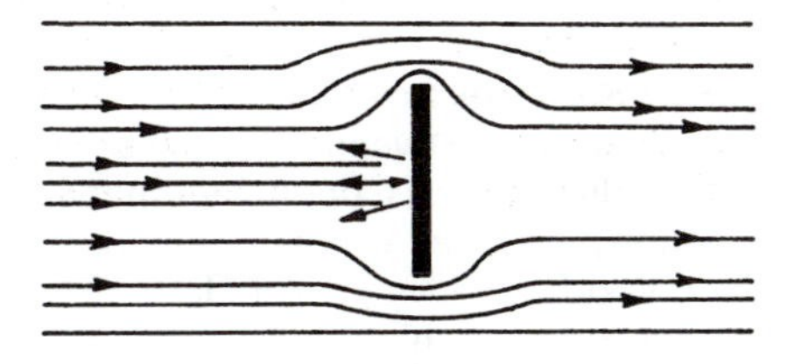

Contours of equal loudness plotted against frequency for pure tones at various intensities. The various curves represent different values in *phons*. The *phon* is a unit that takes into account the unequal sensitivity of the ear. It relates the intensities of sounds at various frequencies to an equally loud sound at 1 kHz. The solid line represents average sensitivities at age 20 years, the dotted line age 60. Note how our sensitivity to low frequency diminishes at low volume. This is the reason for tone-compensated loudness controls. The broken line at the bottom represents minimum levels of audibility.

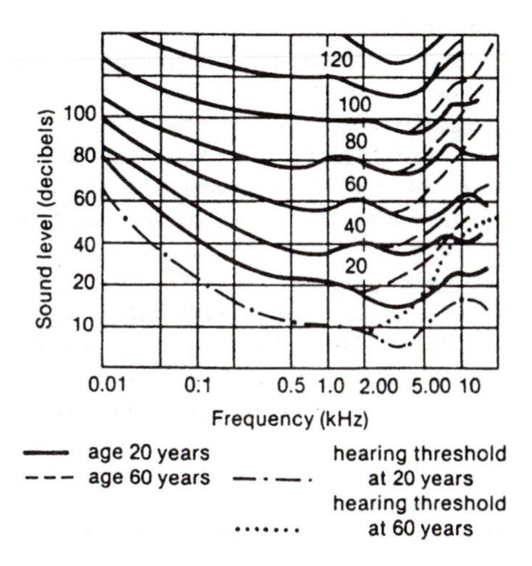

Sound quality is subject to a number of different types of distortion which must be eliminated for high fidelity reproduction.

Sound Quality

The purpose of sound recording is to store information in such a manner that it can be retrieved later. Ideally the signal eventually reproduced should be exactly as recorded.

Frequency response
The recording/reproducing system should have a flat overall response over the desired frequency range – typically 20 Hz to 20 kHz. At first sight (see opposite), it would appear that the extreme high frequency range is unnecessary, especially as the main part of the musical scale and bulk of the audio power is in the middle/lower register. The high frequency range is important, however, as it contains the overtones, or upper partials, which give the various sound sources their characteristic timbre or quality. These overtones can extend to a very high order.

Harmonic structure
The difference between a particular note played on, for example, a piano or a violin depends upon the shape of the resulting waveform. Every waveform, no matter how complex, can be resolved into a fundamental (or lowest frequency), which determines the pitch of the note, and a series of overtones of higher frequency. The latter may be harmonics, i.e. multiples of the fundamental frequency, or unrelated inharmonics such as feature largely in the initial impact of percussion instruments. These starting transients are particularly significant in establishing the timbre, i.e. quality, of an instrument.

Amplitude response
Variations in level of the input signal should be matched throughout by equal variations in the reproduced output otherwise *amplitude distortion* is present. An example of the deliberate use of amplitude distortion occurs in compression amplifiers where the variations in input level result in lesser variations in the output. (See p. 72.)

Linearity
Any non-linearity that occurs in the processing of the waveform, e.g. overloading, causing flattening of the curve, can result in the production of spurious harmonics or *harmonic distortion*.

Intermodulation
If the various frequency components in a signal are caused to intermodulate with each other (multiply instead of add together) due to non-linearity in the system, *beat* frequencies (i.e. sum and difference tones) are produced which can form a long series of harmonics some of which will be discordant. This is known as *intermodulation distortion*. It creates an effect known as *acoustic roughness* which is found to be the most objectionable when the beat rate (frequency difference) is of the order of 50 Hz for frequencies around 3 kHz.

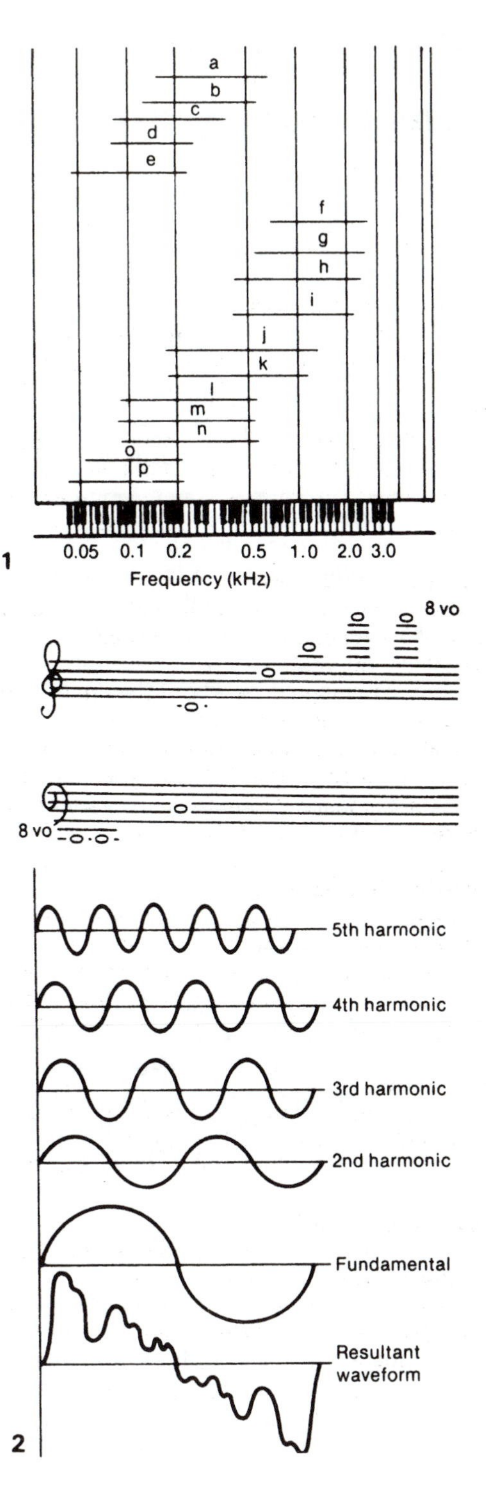

The tonal range of voices and some musical instruments: a: Soprano. b: Contralto. c: Tenor. d: Baritone. e: Bass. f: Piccolo. g: Violin. h: Viola. i: Flute. j: Clarinet. k: Trumpet. l: Bass clarinet. m: Bassoon. n: Cello. o: Tuba. p: Double bass. Only the fundamental frequencies are represented. The overtones extend throughout the upper audio frequency range.

Analysis of a complex waveform into its fundamental and harmonic frequency components. The shape of the final waveform depends not only upon the number and strength of the individual overtones but also upon their relative phase, i.e. their time relationship to each other.

Listening via a microphone and loudspeaker can be very different from hearing the same sound direct.

The Hearing Process

Before considering recording processes in detail it is worthwhile to think about the nature of the material that is to be recorded.

For many people, their first attempt to record with a microphone is the first time that they come up against the problems of acoustics. They make recordings and are surprised to find how different they sound, even when the equipment used is technically immaculate. This can be largely due to the basic difference between listening in person and through a microphone. It is, perhaps, not generally realized that the combination of our ears and our brain provides us with an incredibly complicated mechanism that enables us to hear selectively.

The 'cocktail party' effect

If you are in the middle of a crowd of people, all talking to each other, it is usually quite possible to focus your hearing on a conversation that you find interesting, while effectively pushing all the other sounds into the background. You do not have to turn your head (although this helps) or move your ears like a cat. Your brain interprets the minute differences in time, volume and quality of the sound as it reaches each of your ears in turn. This gives you the ability to judge the direction of the sound sources and discriminate accordingly. On the other hand, microphones, although they may have directional characteristics, are not able to vary them from instant to instant. They pick up everything within their range, including all the reflected sounds that bounce back from the walls, floor and ceiling. Reproducing them in proportion through a loudspeaker robs the listener of his directional ability because all the sound comes from the same place.

Binaural hearing

The ability of the listener to judge direction and discriminate between sources of sound can be restored to some extent with stereophony and almost completely with binaural reproduction (see pp. 76 and 78). There still remain, however, such factors as the volume range normally being greater than is acceptable in the domestic situation and the inability of the eyes to direct the hearing and focus attention.

General factors

In general, the main differences between listening through a microphone and hearing direct are usually:

1. Greater awareness of the acoustic environment.
2. Less ability to distinguish between sources and less clarity.
3. Apparent increase in distance from the source.
4. Greater volume range (unless compression is applied).

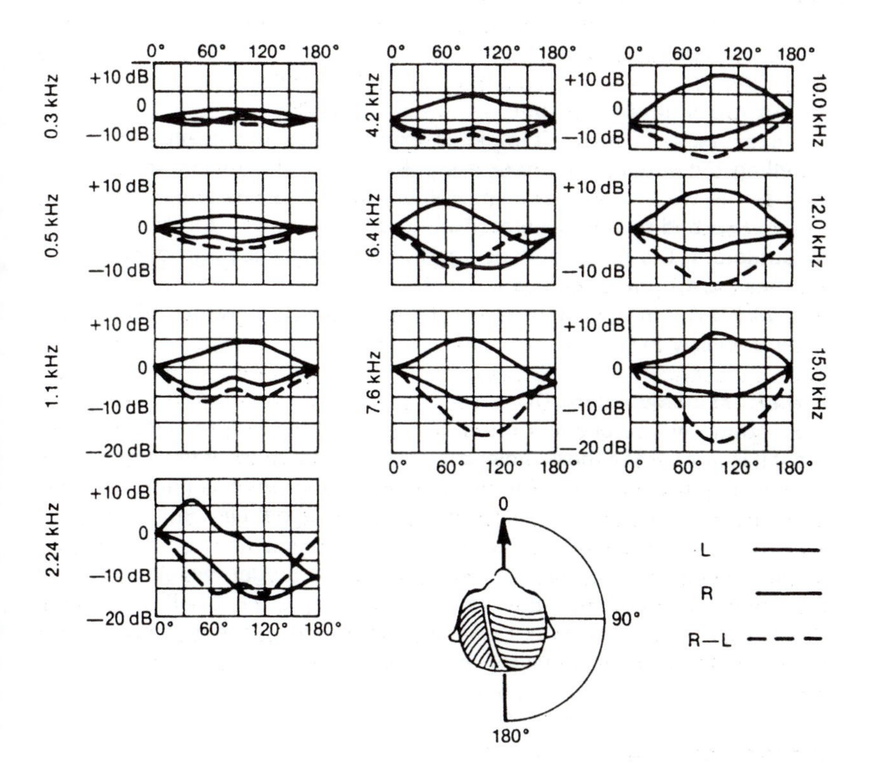

We are able to judge the direction of a sound source because of:

1. The difference in time at which the sound reaches each of our ears.
2. The difference in sound pressure between them.
3. The difference in sound quality due to the acoustic effect of the shape of our ear lobes and head.

The diagram illustrates the way in which the sound reaching each of our ears differs with differing angles of incidence and at different frequencies.
It will be seen that there is little directional effect in the extreme low frequencies, marked difference in level above 6000 Hz and considerable difference in response at high frequencies.

When sound waves hit an object they are either reflected, refracted or absorbed.

Acoustics

On the previous page mention was made of the difference between listening in person and listening through microphones, one aspect of which is the more obvious effect of the acoustics. Let us consider what is meant by acoustics and their effect on the sound.

Reflection and absorption

When a sound is made it radiates from the source in all directions, travelling outwards at approximately 340 m/s until it meets an obstacle such as the walls, floor or ceiling. When it does so it is either wholly or partly absorbed or reflected, according to the nature of the substance it encounters. If the material concerned is porous, dense and of sufficient thickness, the sound waves will penetrate it and use up energy in finding their way through the 'labyrinth' of particles, which therefore vibrate and expend energy in the form of heat.

The effect of wavelength

If the sound waves encounter a substance with a hard polished surface and if it is larger than the wavelength of the sound, most of the energy will be reflected in much the same manner as a mirror reflects light. Because of the very wide range of wavelengths involved and their comparable size with the objects they encounter, however, sound waves tend to suffer considerable diffraction and diffusion. Waves of longer wavelength than the dimensions of an obstacle will curve around it (especially if it has a streamlined shape) creating a small 'sound shadow' immediately behind the object but otherwise having little effect.

The walls, floor and ceiling of most rooms are reflective in their natural state. Plaster reflects about 97% of impinging sound and wood floors approximately 90%. Windows tend to be very reflective but this can be greatly reduced by curtains. The absorption characteristics of draperies depend on the density of the weave and the thickness of the material. Generally speaking this material will only absorb very high frequency sound. To reduce reflections at lower frequencies, draperies must be thick and, preferably, spaced away from the walls. In this way the lower frequencies (longer wavelengths), which would penetrate the material, use up their energy by attempting to make the sluggish mass vibrate.

In the same way, carpets and soft furnishings absorb sound but tend to have more effect at the top end of the frequency scale so that a room which is heavily furnished usually has a dull, 'woolly' acoustic.

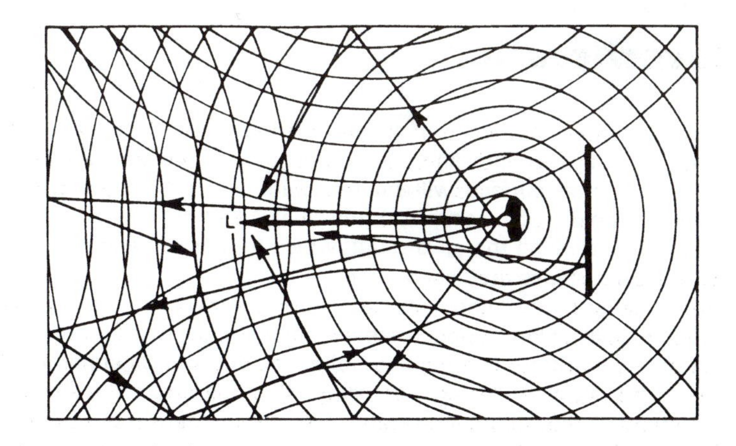

1

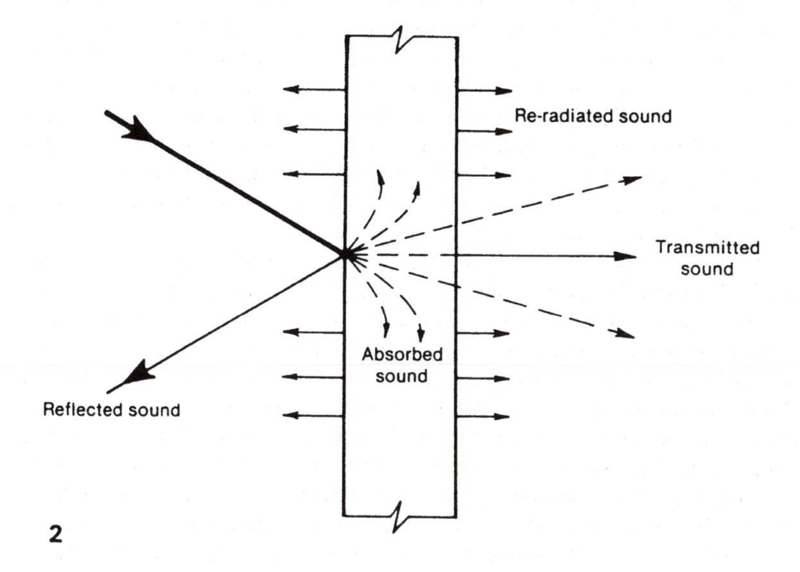

2

1. Sound radiates from the source in all directions. It rebounds from any reflecting surface larger than its wavelength and is refracted around surfaces that are smaller. It meets the listener (L) via a very large number of different paths.
2. When sound strikes a solid object such as a wall, some of the energy will be reflected, some absorbed. Unless the structure is very solid, some will be transmitted through it and some re-radiated by vibrations set up within it.

Sound energy in an enclosed space can build up and be sustained by multiple reflections, called reverberations.

Reverberation

When a sound is produced in an enclosed space, the waves will hit the walls, floors and ceiling and the proportion of their energy that is not absorbed in the surface will be reflected. The reflections will go on bouncing around the room, progressively losing energy on the way through friction with the air through which they travel and absorption at the various surfaces they encounter.

To a listener in the room, especially if listening via a microphone, these reflections will overlap and merge together so that the sound will be reinforced in volume and continue after the source has stopped. This is why people enjoy singing in a bathroom that has hard reflective surfaces (tiles) which provide effective reinforcement to the power of the voice. The overlapping reflections also tend to blur the diction and obscure minor detail in performance.

The prolongation of sound by reflection is called reverberation (sometimes wrongly called echo). It can be plotted for a given room and quantified as *reverberation time* in seconds or fractions of a second. Echo is a single reflection so delayed from the original sound that it is separately identifiable.

Reverberation time

The definition of reverberation time is 'the time that a sound, which has been cut off abruptly, takes to reduce to one millionth of its original power (i.e. to drop through 60 decibels)'.

Typical reverberation times are:

Average living room	0.6 s	Large concert hall	1–2 s
Small talks studio	0.4 s	Large church	5–8 s
Light orchestral studio	0.8 s		

Decay curve

Just as important as the reverberation time is the manner in which the sound decays after it has been cut off. It is normal to assess studio acoustics by making an impulsive noise (e.g. a pistol shot) and plotting the shape of the decay curve with an oscilloscope measuring volume level against time. A good recording venue would have a smooth decay curve, reasonably equal throughout the frequency range, parabolic in shape and of appropriate length. Any humps in the curve would suggest standing wave reflections (eigentones) or ringing in the structure. These are to be avoided because they will produce a distorted, unclear recording.

In a small room standing wave effects can be mitigated by placing absorbent material between any parallel surfaces and not using microphones at the focus of reflective surfaces.

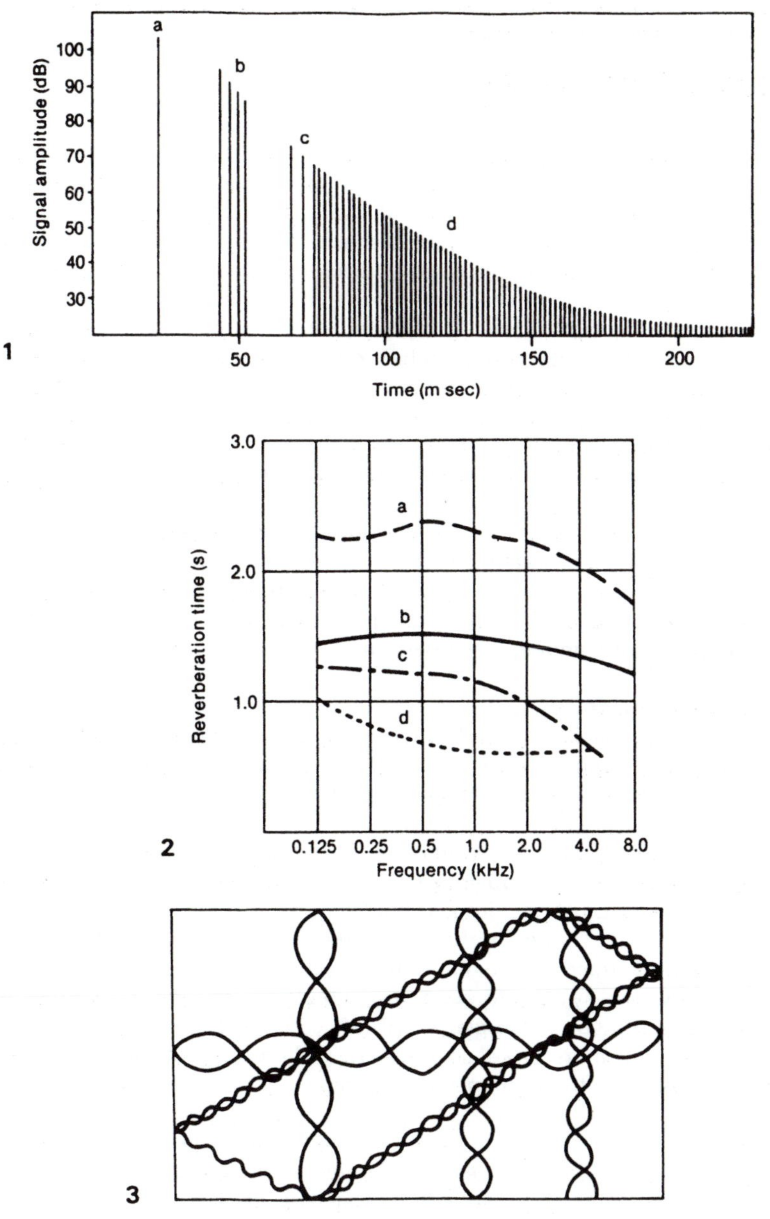

1. Illustrating the response of an auditorium to an impulse sound: (a) the initial sound; (b) the first few reflections from comparatively close reflecting surfaces; (c) reflections from nearest walls; (d) multiple reflections from the body of the hall.

2. Typical reverberation times for various types of room and for various frequencies. (a) Very large hall; (b) concert hall; (c) large lecture room; (d) living room.

3. Some standing wave patterns (eigentones) in a room. These cause resonances and colouration in a small room.

The first requirement to produce a good recording is to choose, or arrange for, the correct acoustic environment.

Choosing the Right Acoustic

When it comes to assessing the suitability of a room or hall for recording it is important to relate the reverberation time (p. 22) to the size of the room and the type of programme material to be played. A long reverberation time, which would do much to enhance the blending of tones in well balanced classical music, could inhibit the acoustic/microphone separation necessary to achieve the right effect with a light orchestra that is not internally balanced. It could be quite useless for speech because the overlapping reflections could obscure the diction and make it unintelligible.

Acoustic compromise

Most recording studios have to accommodate a variety of different musical combinations and it is not unusual for the room to be much smaller than the ideal. It is, therefore, generally preferable to err on the side of too little reverberation in the studio. This will enable a number of microphones to be used with the minimum of mutual interference and will mitigate the 'boxy' effect that can be produced in a small room through the formation of standing wave patterns.

Performer comfort

The room must not be so 'dead' that it is oppressive to perform in. Anyone who has been in an anechoic chamber (a room with walls totally lacking in reflection used for testing microphones) will know just how oppressive this can be.

Where a quiet artist is required to perform in a 'dead' acoustic or with a loud accompaniment (an obvious example is a vocalist with a 'pop' group backing), it may be almost impossible for him to hear himself so that his performance could suffer. In these circumstances it is usual to 'fold back' a portion of the microphone output to a loudspeaker close by, taking care not to allow the system to become unstable (howl-round), or provide headphones. However, the placing of reflective acoustic screens around individual performers or sections can help them to hear themselves as well as improving the sectional separation.

Noise

A very important factor in the choice of venue for recording is insulation from extraneous noise. The tolerable noise level depends upon the nature of the recording and the closeness of the microphone technique.

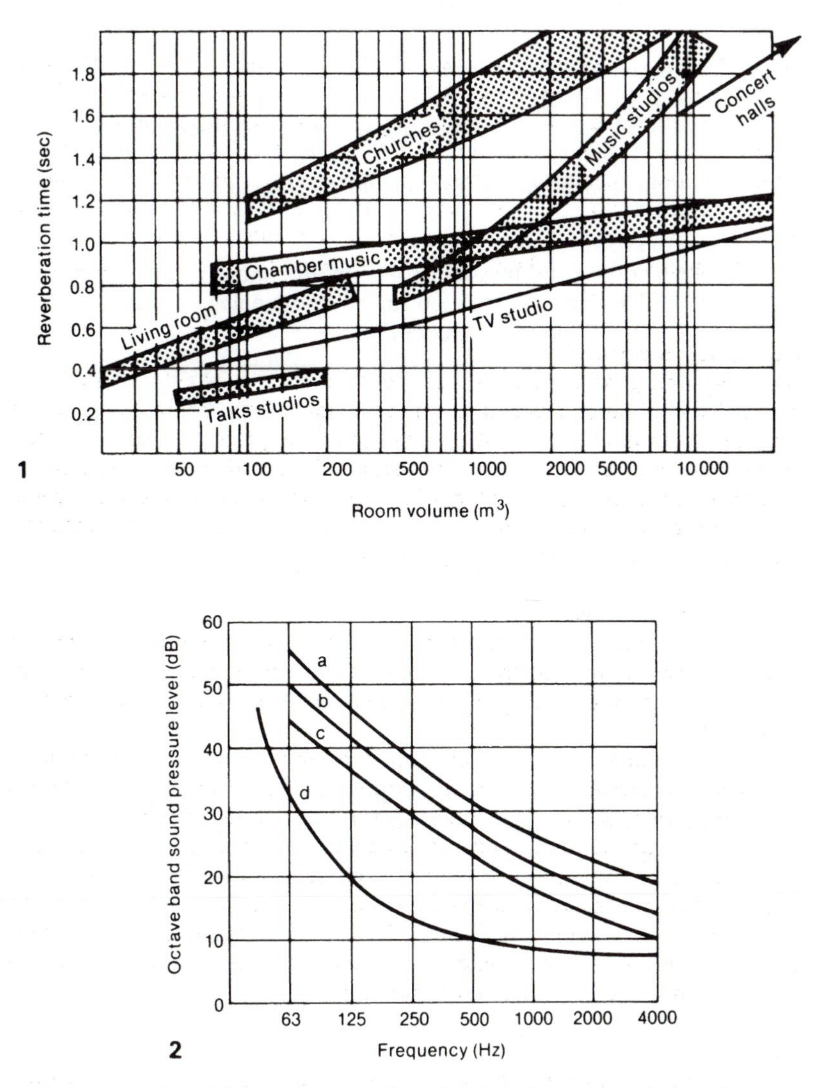

1. Recommended mid-frequency reverberation time for various sizes and functions of rooms.

2. Recording venues should be as quiet as possible. The table gives the accepted criteria for background noise levels in broadcasting studios: (a) audience studios; (b) television and sound studios (except drama); (c) sound drama studios; (d) quiet sound studios.

Rooms to be used regularly for recording should have their acoustics measured and adjusted.

Acoustic Treatment

If a room is to be used only occasionally for recording, and particularly if a measure of acoustic variation is required, possibly the best arrangement is to provide a thick, felted carpet for the floors and thick velour drapes which can be drawn to cover any windows and at least two adjacent walls. These should be spaced about 10 cm (4 in) off the walls.

Acoustic measurement

A room that is to be used frequently as a studio should, ideally, have its reverberation characteristics plotted for all frequencies and have proper acoustic treatment designed to match.

Absorption coefficient

The reverberation time can be measured by making a loud noise and measuring the time the sound takes to decay through 60 dB (see p. 22). Alternatively it can be assessed theoretically from a knowledge of the absorption coefficients of the various materials used in its construction. The absorption coefficient (α) of a material is the proportion of the incident energy arriving from all directions that is not reflected back. The following are some examples of midfrequency absorption coefficients, α:

plastered wall	0.03
wood floors	0.1
thick pile carpet	0.5
curtains (medium) hung in folds	0.6

Sabine's formula

A fair approximation of the likely reverberation time of a room can be assessed by working out the mean absorption coefficient from the relative areas of the various absorbent materials and applying Sabine's formula:

$$\text{Reverberation time} = \frac{0.164V}{S\alpha}$$

where V is the room volume (m^3)
S is the room surface area (m^2)
α is the mean absorption coefficient

The desirable reverberation time depends upon the size of the room and the purpose for which it is required.

Acoustics can be adjusted by panel absorption units. The most useful type is the wide-band absorber that combines porous absorption for high frequencies with membrane absorption for the bass. Construction of a wide-band absorber:
(a) perforated hardboard; (b) bituminous roofing felt, bonded to the back of the hardboard; (c) rockwool or glass fibre; (d) timber frame; (e) wall.
The rockwool or glass fibre absorbs the high frequency sounds which penetrate the holes and the roofing felt. The roofing felt damps the hardboard due to its inherent sluggishness and causes it to act as a damped membrane absorber. A wide range of absorptions can be arranged by making up boxes of differing sizes and thicknesses.

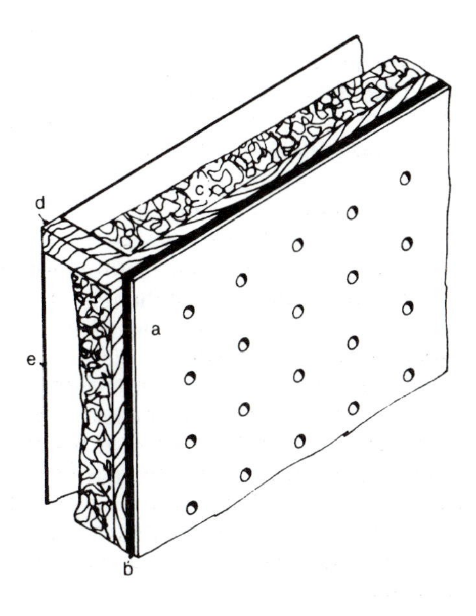

Typical absorption coefficients of wide-band absorbers. The high frequency absorption can be determined by the ratio of perforation of the hardboard. The holes act as cavity resonators (Helmholtz resonators) with the air space behind, for (a) 25%, (b) 5% and (c) 0.5% perforation.

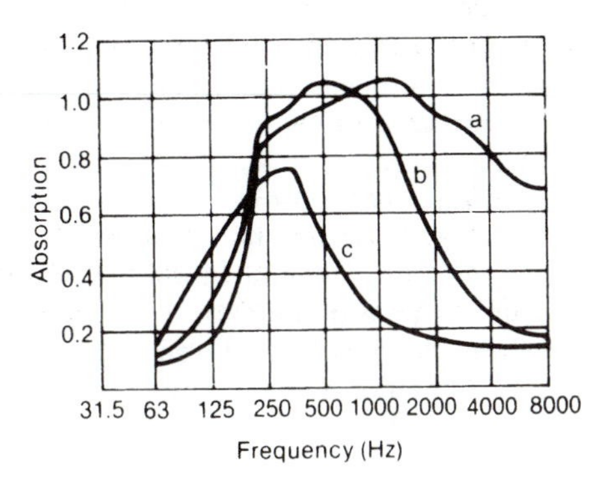

Acoustic Isolation

A measure of the suitability of an area for sound production is the separation that it is possible to achieve between the various sources of sound, both wanted and unwanted.

Noise weighting

The definition of noise is 'unwanted sound'. Although it would be possible to measure the actual sound pressures involved, what is more important is to establish its nuisance value. To assess the effect it will have on the listener it is necessary to take into consideration the unequal sensitivity of our hearing and the way it changes with different sound levels. For this reason noise level meters are usually provided with three scales of 'weighting' corresponding to different phon contours (i.e. curves of equal sensitivity for different sound levels). Three switchable weighting networks are used according to the level of sound corresponding to the 40, 70 and 80 phon contours. The results are said to be A, B or C weighted.

In assessing the suitability of a room a series of *Noise Criterion* curves have been established. The generally accepted standard for a small general-purpose studio is NC 15–20, rather less where the noise is easily identifiable or irritating in character. Small listening rooms should be better than NC 30.

Masking

Where noise is continuous it is important to take into account the masking effect whereby our hearing sensitivity is reduced for frequencies in the presence of other sounds of similar frequency. This is the basis of the Dolby noise reduction system (see pp. 120–127).

Domestic situations

In normal domestic rooms noise tends to be air-borne (mainly through doors and windows) and conducted through the structure. Exterior noise can be reduced considerably by providing secondary double-glazing with the panes spaced 10–15 cm (4–6 in) apart and preferably not quite parallel to each other. Doors should be fitted with draught seals (ideally of the magnetic type as used in refrigerators) and cushioned thresholds. If possible double doors should be provided, forming a sound lobby, but otherwise the door could be covered by a thick drape. It may be necessary to carpet the room above and corridors outside to prevent the transmission of footfalls owing to sound conduction.

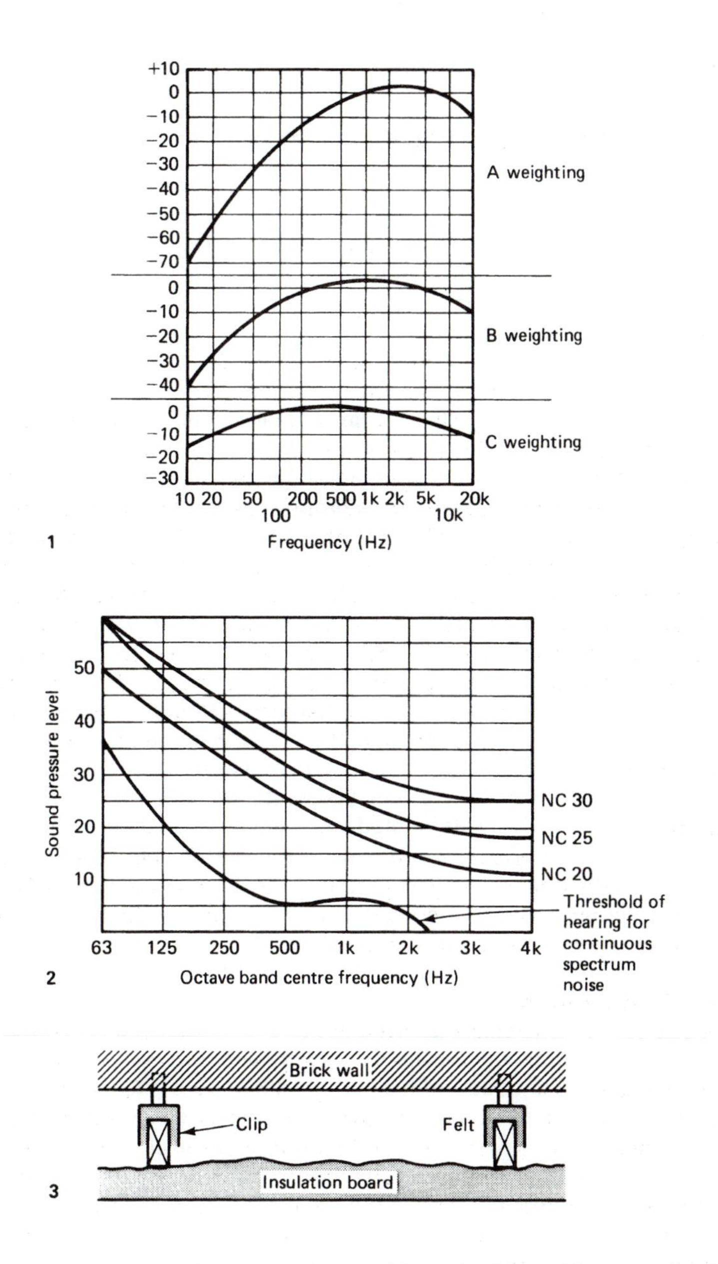

1. The response of weighting networks used in conjunction with meters for the measurement of noise. They are chosen according to the loudness level of the noise and take account of the variation in the sensitivity of our hearing at different intensity levels.

2. Noise criterion curves. Small studios should have a measured NC of 15–20. Listening rooms should be better than NC 30.

3. A simple way to improve insulation from noise through a wall by using insulation board fixed on battens and retained to the wall by felt-lined clips.

Artificial Reverberation

It is unlikely that the acoustics of the room or studio will be ideal; in any case, with modern microphone technique it is normal to apply differing amounts of reverberation to different sections of an ensemble. There is much to be gained by starting with a studio with too little reverberation and adding it artificially where required.

Echo room

The simplest way is to place a loudspeaker and microphone in a reverberant room, feed a proportion of the required microphone output into the loudspeaker in the reverberation chamber (usually erroneously called the echo room) and then add a proportion of the reverberant sound picked up on the microphone back into the main output. The chamber should have reflective walls, ideally not parallel to each other otherwise standing waves will be formed. The path length between loudspeaker and microphone should be made as long as possible by the use of baffles.

Electronic reverberators

It is now most usual and cost-effective to simulate reverberation electronically. Early methods included the 'bucket-brigade' delay line, in which a series of capacitors was charged in succession through a series of resistors, so that the signal was delayed as it passed along the line, each stage being progressively attenuated and added to the output. With digital systems the action of RC networks can be simulated by digital filters. They can be used in cascade with each stage providing delay and adding exponentially decreasing samples to the output. Infinite impulse (recursive) filters have a proportion of their output returned to their input, and by continually returning half the value of each output sample to the input (a form of positive feedback) an exponential response can be obtained.

In fact a recursive arrangement, with a perfect exponential decay characteristic, sounds unrealistic, as real life acoustic usually introduces coloration and an unequal frequency response. So digital reverberators usually provide means of deliberately introducing distortion: comb filters with different delays are added in parallel to cause phasing effects to simulate reverberation in natural environments, or to suit particular types of programme material (see p. 32).

Reverberation plate

A convenient method of obtaining artificial reverberation consists of a steel sheet approximately 2.5 × 1.5 m (8 × 5 ft) suspended vertically on springs. A moving coil transducer is fixed at a critical point near the centre and a piezo-electric contact microphone is mounted near one edge. When a signal is applied to the transducer a series of very complex vibrations radiate out to the sides and are reflected to and fro between the edges until they die out in the manner of reverberation. The reverberation time

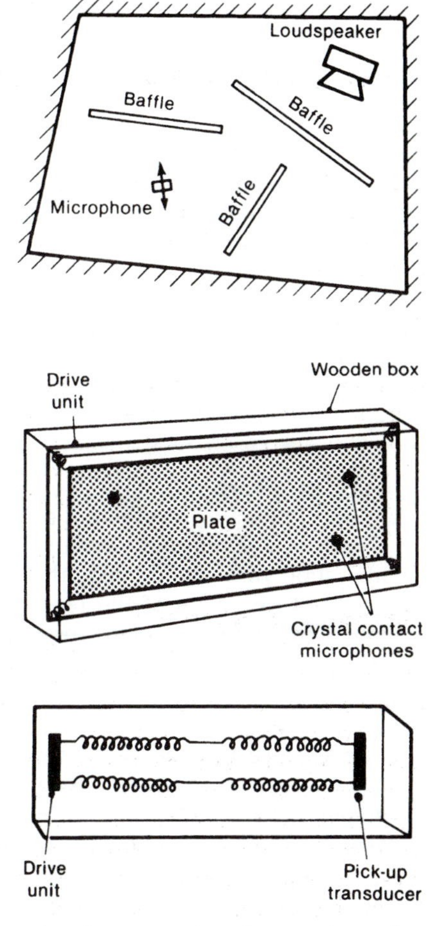

1. Plan of typical arrangement of a reverberation chamber (echo room). If possible the walls should not be parallel to each other. The baffles create a longer initial path between loudspeaker and microphones and help to break up standing wave reflections.

2. Reverberation plate. The specially selected metal sheet is suspended by springs at each corner in a wooden box. Vibrations are induced into it by a moving coil transducer and picked up by one, two (in the case of stereo) or four (for quadraphony) ceramic contact microphones.

3. Reverberation springs. Usually arranged in pairs, each with two springs of different lengths joined together.

can be adjusted by altering the spacing between the plate and another plate of similar size made of porous material held parallel to the first. The second plate effectively damps the vibrations in the reverberation plate.

Reverberation springs

The most compact mechanical device consists of two sets of springs of different lengths joined together at the ends. Sound waves are induced into one end of the springs in a torsional manner and received at the other. The energy reflects to and fro between the ends of the springs until it dies away, giving a reverberant effect. Another design uses a torsionally driven spring, providing a decay time of 2–4.5 s.

Time delay

Whichever reverberation device is used it is likely to be more natural if the sound is delayed before the reverb. This is because, in the natural situation, the sound waves have to travel to the walls and back to the listener (at about 1 ms per foot) before the first reflections take effect.

Using Reverb

Unless the natural acoustic conditions happen to be perfect, most sounds can be improved by starting with too dry an acoustic, or with close microphone technique, and adding artificial reverberation of a suitable character. The various parameters are as follows.

Pre-delay
This represents the time that, in the natural situation, the sound reflections would take to build up before reverberation becomes effective. It is related to the size of the room. Early reflections can enhance presence and impact.

Amount of reverb
The ratio of direct to reverberant sound (echo mixture), this determines the apparent volume and warmth of the sound. We tend to associate reverberation with sound 'filling the hall'. (If you whisper in a large hall you will not be aware of the reverberation. Shout and it will be very obvious. The reverb time of the room does not change, just the amount of the reverb that is loud enough for you to hear.)

Rate of decay
This determines the apparent size of the room. A slow rate of decay suggests a large room (about 2 seconds for a concert hall, 5 seconds for a cathedral). A large proportion of reverberant sound with a rapid decay would suggest a small, very reverberant room (e.g. a bathroom). This type of reverberation can be used to 'fatten' the sound, give the effect of power, particularly in the bass, and cover imperfections. (It is encouraging to sing in the bath.)

Reverberation frequency response
Analysis of various studios and concert halls shows that the best music studios have reverb frequency responses that are far from flat. Usually they have a marked, but smooth, increase in the 200 Hz to 2 kHz region. This gives a warmth and sonority to the sound. The frequency responses of reverb devices should be similarly adjusted to suit the character of the sound.

Diffusion
Digital reverb devices provide the ability to adjust diffusion, i.e. the smoothness of the decay. Natural acoustics are usually best when the diffusion adopts a smooth exponential curve. However, this may not always be ideal, for example if reverb is being used in a particular channel to 'fatten' the sound, say, of a drum. Low diffusion results in a grainy sound which can suit vocals.

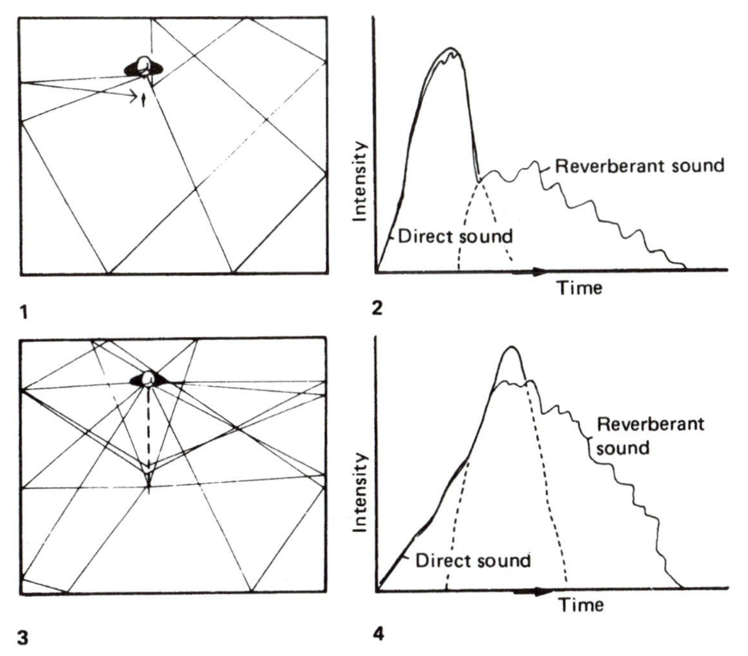

1. Artist working close to microphone. The direct sound reaches the microphone well in advance of the reverberation, which is at much lower volume.

2. The sort of decay curve to be expected from an artist making an impulsive sound at close range in a reverberant room. Our ears can recognize that the direct precedes the indirect sound, and this gives us the clue that the sound is close, but the room is reverberant.

3. Artist at some distance from the microphone. The reflections arrive at a similar time to the direct sound, which appears to be submerged in the reverberation.

4. Slower build-up and later decay of sound. Our ears associate this effect with distance.

Depth

The apparent distance of a sound source is a function of pre-delay, ratio of direct to reverb sound and decay slope.

Chorus

Digital reverb devices can offer effects such as multiple repeats with slightly delayed timing and pitch changes (phase) to give the effect of multiplying the number of musicians and thickening the sound.

Flanging

If the delay period is dynamically changed between zero and about 10 ms the result is 'flanging', an effect where a notch sweeps through the audio spectrum. (Originally obtained by adding the outputs of two tape recorders which are recording the same signal, and varying the speed of one of them.)

Sound recording begins with a microphone, a device for converting acoustic energy into mechanical or electrical power.

Microphones

The aim of sound recording is to store mechanical energy derived from the minute movement of air particles caused by sound waves in a form that can be retrieved later. In modern practice the process begins with the use of microphones.

A microphone is a transducer. It can convert acoustic energy into electrical power, either by direct contact with the vibrating source or, more usually, by intercepting sound waves radiating from a source.

Microphones that pick up sound from waves in the air have diaphragms to collect the acoustic energy. They fall into two basic categories according to whether one or both sides of the diaphragm are exposed to the sound waves.

Pressure microphones

Pressure-operated microphones have only one side of their diaphragm exposed to the sound waves; the other side is sealed off except for a small aperture to allow the atmospheric pressure to equalize on both sides on a long-term basis (like the Eustachian tube in the ear). These microphones respond to minute variations in the surrounding air pressure. They therefore operate irrespective of the direction of the sound source unless their physical characteristics dictate otherwise (e.g. their dimensions are comparable with the wavelength of the sound) and are said to be omnidirectional.

Pressure gradient microphones

Microphones that have both sides of their diaphragms exposed to the sound waves operate by virtue of the *difference* in pressure (i.e. the pressure gradient) existing between the front and the back at any moment in time. This difference is a function of the different path length for the sound waves between the front and the back of the diaphragm. Obviously this path length difference will vary with the angle of incidence; this type of microphone, therefore, has a directional response.

Microphone types

There are four main types of microphones (discounting specialist types, such as thermal and ionic transducers, which are normally only used for scientific purposes). They are:

1. Carbon granule, which exploit the variation in resistance with pressure of a loose-fill of carbon granules. These were used extensively in telephones.
2. Piezo-electric, which use the property of some crystals to generate small electrical potentials when physically stressed by sound pressure variations.
3. Dynamic (moving-coil) and ribbon.
4. Capacitor (condenser) microphones.

Types 1 and 2 are virtually obsolete and will not be discussed here.

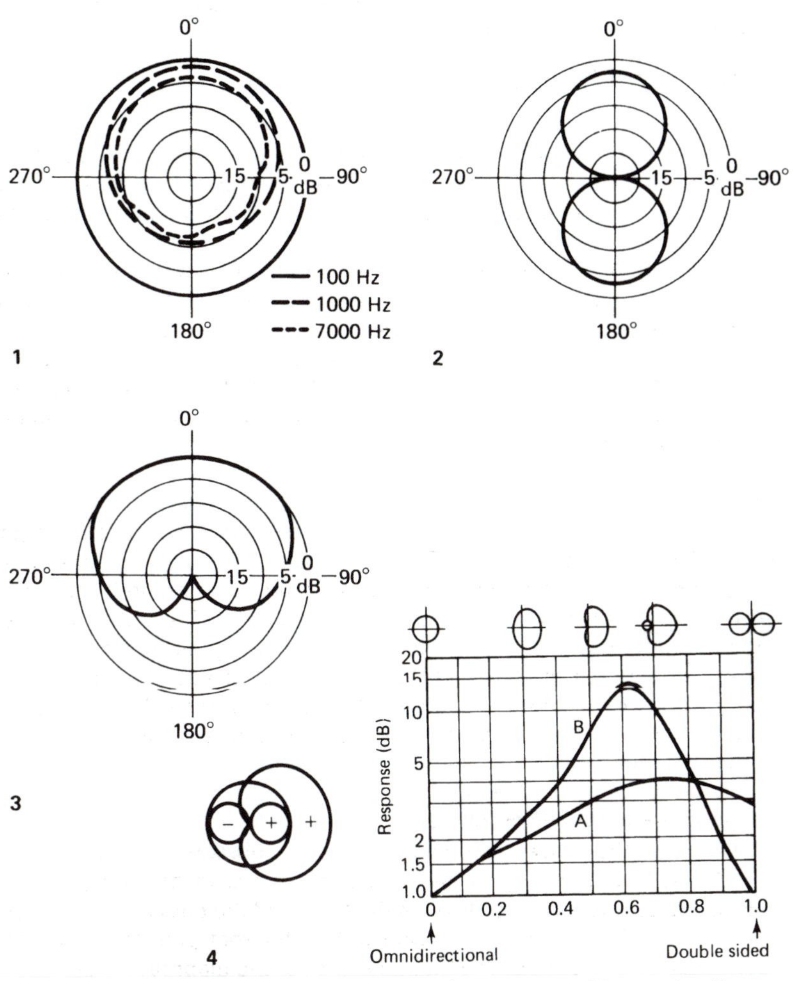

The directional response of microphones can be represented by a polar diagram. This is a circular graph in which the centre represents the microphone. The distance from the centre of a line drawn around it represents the sensitivity at each angle with respect to its front axis.

1. A perfectly omnidirectional microphone would have a circular polar diagram. In the one depicted this is only true for low frequencies (100 Hz). At higher frequencies the 'shadow' of the microphone case reduces the output.

2. The figure-of-eight characteristic is typical of a ribbon microphone.

3. Cardioid (heart-shaped) polar diagram. The cardioid shape can be obtained by adding together equal omnidirectional and figure-of-eight outputs.

4. Limacon curves show the effect of adding together different proportions of omnidirectional and figure-of-eight outputs.
Curve A gives the result in terms of directivity, i.e. the ratio of incident to ambient sound. Curve B shows the unidirectional factor, i.e. the ratio of axial response to response from all other directions.

Moving-coil microphones are robust, can have omnidirectional or unidirectional response and are of good quality.

Moving-coil microphones

The moving-coil microphone consists of a diaphragm, which is usually a circular disc of aluminium alloy or plastic formed into a shallow dome with corrugations around the edge, close to where it is clamped, to allow it to move freely. To the back of the diaphragm is attached a coil of wire, usually wound from flat aluminium wire for lightness, which is suspended by the diaphragm in the annular gap between a powerful magnet and a central pole piece.

Induced voltage
When sound waves impinge on the front of the diaphragm, the back of which is partially or completely sealed off, the changing air pressure causes it to move in and out. This causes the coil, attached to the back of it, to move in the magnetic field, cutting the magnetic lines of force at right angles. As explained on p. 142, this induces an electromotive force, which constitutes the output, in the coil.

Microphone impedance
In order to reduce unwanted resonance of the diaphragm, the coil must be kept light and therefore consists of relatively few turns. Consequently the impedance of moving-coil microphones tends to be low, characteristically about 30 Ω. Some incorporate a small step-up transformer into the stem of the microphone to increase the impedance to about 200 Ω. Some moving-coil microphones have impedances that rise with frequency and need to be connected to amplifiers with an input impedance of about five times their own impedance to obtain a flat frequency response.

Omnidirectional dynamic microphones
Dynamic microphones that have the back of their diaphragms sealed from the air (except for a small pressure-equalizing aperture) have an omnidirectional response for all frequencies with a wavelength longer than their dimensions. The larger types of microphone can create a 'sound shadow' which reduces the high frequency response off-axis.

Directional dynamic microphones
Dynamic microphones can be made directional by allowing a proportion of the incident sound wave to reach the back of the diaphragm through some form of acoustic delay. This can consist of a tube to increase the path length or some form of acoustic labyrinth such as a fine porous material. The acoustic delay produces a difference in the relative pressures between front and back of the diaphragm which varies with the angle of incidence. Thus, by careful design, a cardioid (single-sided) response can be obtained.

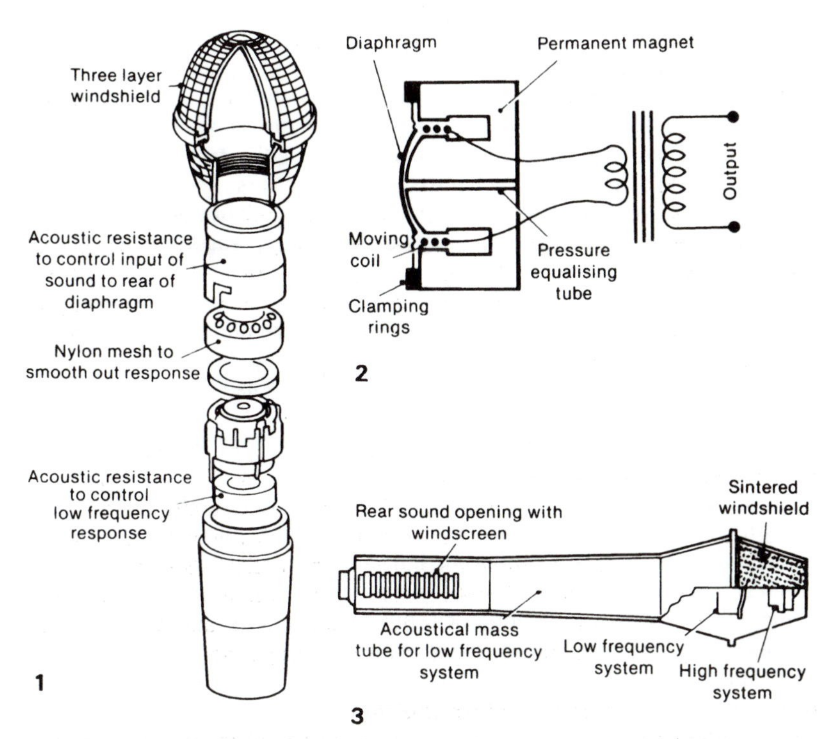

1. Sectional view of a directional moving-coil microphone. Sound waves approaching from the front axis of the microphone reach the front of the diaphragm direct and the back via an acoustic delay so that a phase difference is created. Waves from the back encounter the acoustic resistance first, arrive at both sides of the diaphragm in the same phase and cancel.

2. Simplified diagram of the construction of a moving-coil microphone. The diaphragm is dome shaped to promote stiffness.

3. A double element directional microphone. The problem of maintaining the proper phase relationship to achieve a cardioid response throughout the frequency spectrum is alleviated by dividing it into two with a separate transducer for each: one for the range 20–400 Hz and one for 400–18 000 Hz.

Ribbon microphones can have excellent response characteristics but are prone to handling noise and wind blasting.

Ribbon Microphones

In the ribbon microphone the diaphragm consists of a narrow ribbon of corrugated metal foil stretched between the pole pieces of a powerful magnet.

Pressure gradient operation

The ribbon is exposed to the sound waves on both sides. It is activated by the difference in sound pressure between the front and the back of the ribbon at any instant in time. Sound waves approaching from the front of the microphone take longer to reach the back and vice versa so that a difference in pressure is created between the front and the back of the ribbon, which causes it to move away from the direction of higher pressure.

Electromagnetic induction

As in the moving-coil microphone (p. 36) the movement of the conductor (in this case the ribbon) in a magnetic field induces an electric current into it. This is applied to a step-up transformer (to increase the impedance) to provide the output.

Directional characteristics

Operation of the ribbon microphone depends upon the difference in path length between front and back, so it will have maximum sensitivity at the front and back and zero output at right angles, where the path length to each side of the ribbon is identical. In a polar response curve, in which the sensitivity of the microphone in any direction is represented by the distance of a line from a point representing the microphone, the microphone will have a figure-of-eight configuration. It can be considered to be live over an angle of approximately 90° at the front and back and dead at the sides.

Ribbon microphones have figure-of-eight polar characteristics in both the vertical and horizontal planes but they tend to be lacking in HF response off-axis in the vertical plane when the wavelength of the sound becomes comparable with the length of the ribbon, causing a wave motion along its length.

Unidirectional ribbon microphones

Ribbon microphones can be given unidirectional characteristics by blocking off the back of the ribbon with an acoustic delay. This matches the two path lengths so that sound approaching from the rear reaches the front and the back (through the delay) simultaneously and cancels out.

General characteristics

Ribbon microphones can have excellent frequency response characteristics but they have a small, low impedance output, tend to be delicate and suffer from handling noise and wind blasting.

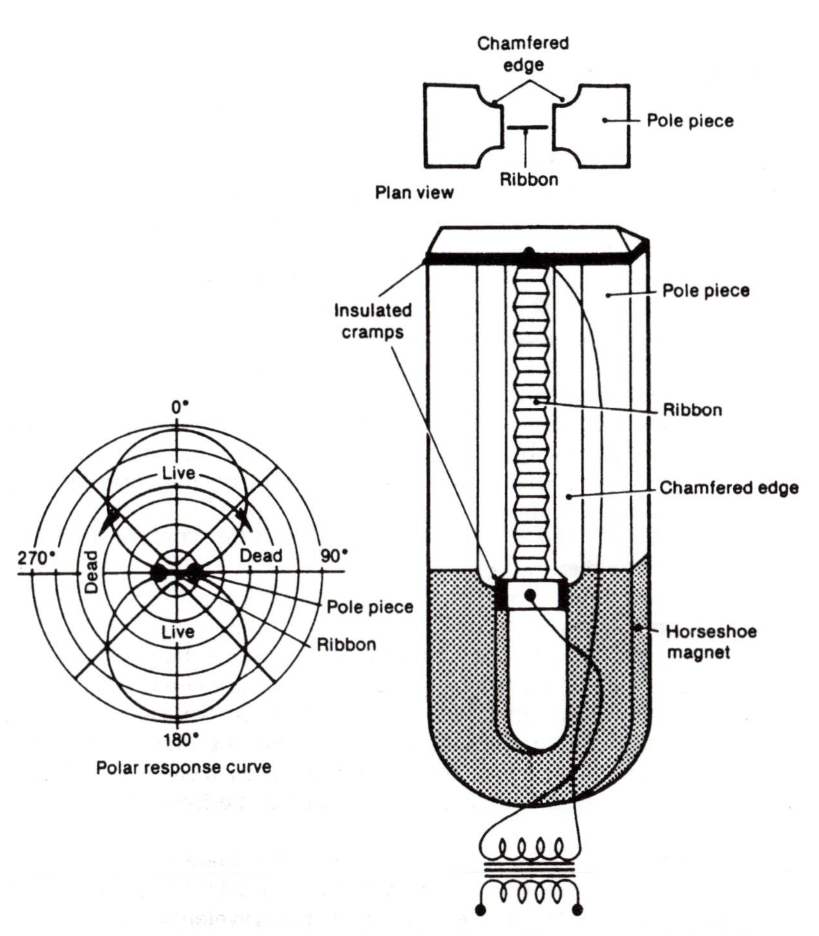

Polar response curve

The ribbon of a ribbon microphone is made of very thin corrugated aluminium alloy. It is held by two insulated cramps between the pole pieces, which are extensions of the powerful horseshoe magnet.
Sound has equal access to both sides of the ribbon and high frequencies are also able to pass between the gaps at the sides of the ribbon. To offset the resulting loss the pole pieces are usually chamfered. This creates an acoustic resonance which boosts the HF response.
Ribbon microphones have figure-of-eight polar characteristics in both the horizontal and vertical planes but they tend to be lacking in HF response off axis in the vertical plane when the wavelength of the sound becomes comparable with the length of the ribbon, causing a wave motion along its length.

Condenser Microphones

The condenser microphone is really a small variable capacitor in which one plate is formed by the backplate and the other by the diaphragm, which is evenly spaced very close to it. Characteristically, the distance between the plates is about 25–40 micrometres (1–1.5 thousandths of an inch) giving a capacitance of about 10–20 pF.

Variations in sound pressure cause the diaphragm to move with respect to the backplate so that the capacitance varies in relation to the sound wave.

Polarizing the capacitor

In order to obtain an output the variable capacitance must be turned into an emf. This is done by applying a polarizing potential of about 50 V through a high resistance (100 MΩ or more). Variations in capacitance caused by movement of the diaphragm are thus converted into voltages across the resistor, which are then applied to a head amplifier (usually consisting of a field effect transistor) contained in the stem of the microphone. Some condenser microphones are provided with co-axial extension tubes which can be interposed between the capacitor capsule and the head amplifier.

Electret diaphragms

Some condenser microphones eliminate the need for a polarizing voltage by having a permanent electrostatic charge, either in the diaphragm or the backplate (rather as a permanent magnet retains its magnetism). This results in a much more compact assembly, especially in the case of miniature clip-on microphones. Although a power supply is still required for the head amplifier this can be separated from the microphone capsule by a short lead.

RF capacitor microphones

Another type of capacitor microphone uses the variation in capacitance to beat with an RF oscillator, the rectified signal comprising the audio frequency output. This type of microphone can have an excellent frequency response and good sensitivity but can produce variable results when subjected to large variations in temperature and humidity.

Directional response

Capacitor microphones can be operated by pressure and are therefore omnidirectional if the back of the diaphragm is sealed (except for a small pressure equalizing path). They can be made into very small capsules which do not produce sound shadows and thus retain their all-around response at all frequencies.

Alternatively the backplates can be made of a porous material which forms an acoustic delay path and produces a unidirectional response.

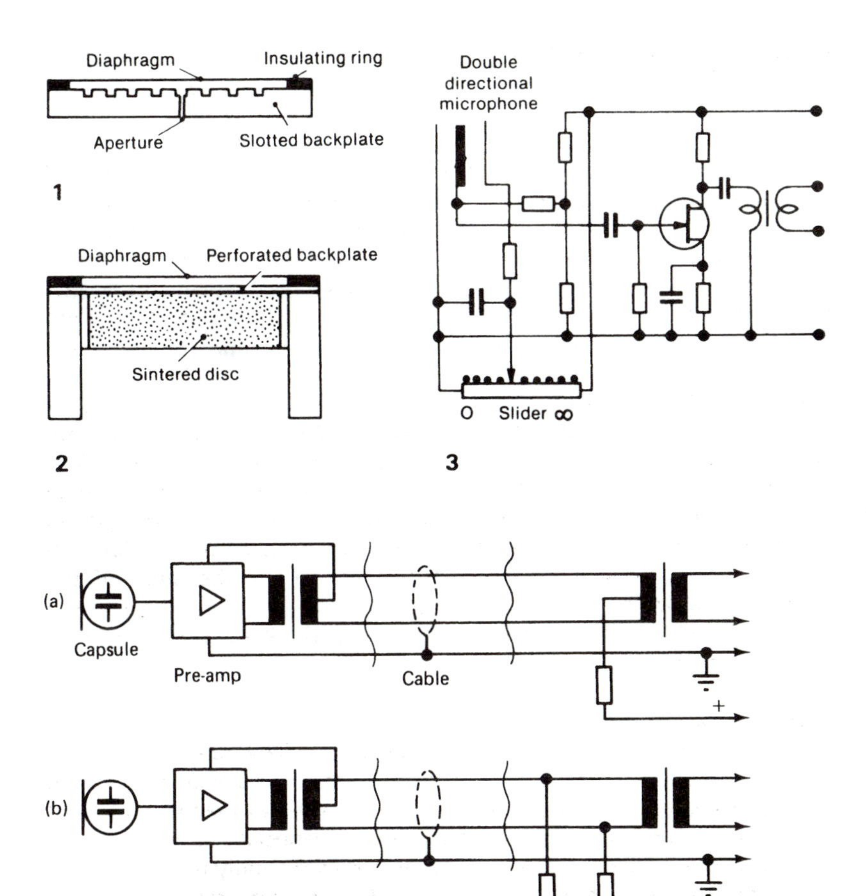

1. Sectional view of a pressure-operated capacitor microphone. The solid backplate is slotted to reduce the back pressure behind the diaphragm. The diaphragm is made of thin aluminium or metallized plastic stretched over an insulating ring. There is a small aperture to allow the pressure to equalize on a long-term basis.

2. Sectional representation of the essentials of a pressure-gradient (directional) capacitor microphone elements. The porous disc forms an acoustic delay.

3. A method of achieving variable directivity using a double-directional microphone, i.e. two diaphragms separated by a shared porous back plate. Moving the slider determines the relative polarity between the plates and produces a range of directivity patterns continuously variable between omnidirectional and figure-of-eight patterns.

4. Two types of phantom powering for condenser mics, to enable the head-amp to be powered via the microphone cable: (a) centre tapped transformer with common resistor, (b) artificial centre tap with two resistors.

Some microphones are especially suited to specific applications.

Vocalist's Microphones

Microphones are available with characteristics especially suited for vocal work, but this is not to say that they may not have many other applications for which their design makes them appropriate.

Close-working vocalist's microphones

On page 46 we note the bass tip-up effect which occurs when a pressure gradient microphone is used close to a point source of sound. A vocalist's mouth can be considered to be a point source of sound, located about 12 mm (½ in) behind the lips, so that the rise in low frequency response can be considerable when the microphone is only a short distance away, except in the case of some double-element microphones where the bass receptor gets its input from the rear of the microphone.

The close-up bass enhancement can be utilized in several ways. Vocalists can use the bass boost to extend their range for low notes, or the bass response can be equalized and the resulting loss for distant sounds used to improve discrimination against ambient sounds and feedback from loudspeakers etc. For this reason many vocalist's microphones are provided with a switch on their case giving such options as:

1. Full sensitivity, full frequency response.
2. Full sensitivity with cut-off below 100 Hz with about 12 dB/octave slope.
3. −14 dB sensitivity, full frequency response.
4. −14 dB sensitivity with roll-off of about 6 dB/octave below 500 Hz.

Some microphones also provide the option of 'presence boost' giving a peak of +2 or +4 dB at 4000 Hz, which accentuates the sibilants and tends to give the impression of closeness. These controls can be used to suit the microphone to the individual characteristics and power of the vocalist or instrument.

Vocalist's microphones normally have cardioid or hypercardioid characteristics, the latter being especially suitable for guarding against feedback from monitor loudspeakers (sited on the floor to help the artist to hear himself) and PA.

Hand-held microphones

Microphones for hand-holding must be robust. Some are shock-mounted within their cases and incorporate double breath screens. One type employs two transducers. One picks up the sound and the other is sensitive only to impact and shock noises, which it combines with the other in antiphase to cancel them out.

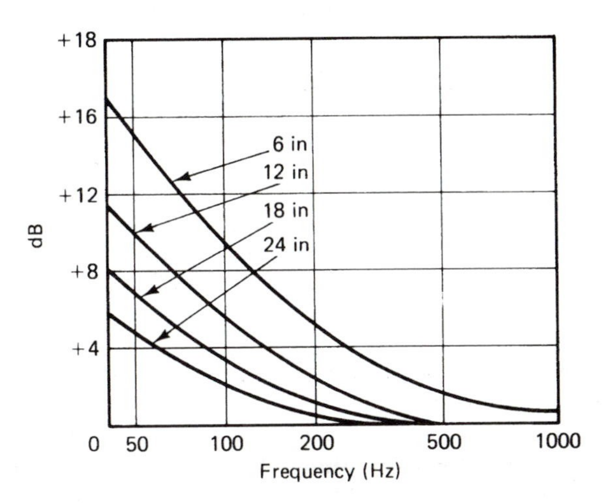

1. Graph illustrating the increase in bass response of a microphone as the distance from the mouth decreases. Even with the microphone 15 cm (6 in) from the mouth (which is greater than the distance usually employed by pop vocalists) the bass is increased by about 16 dB at 50 Hz.

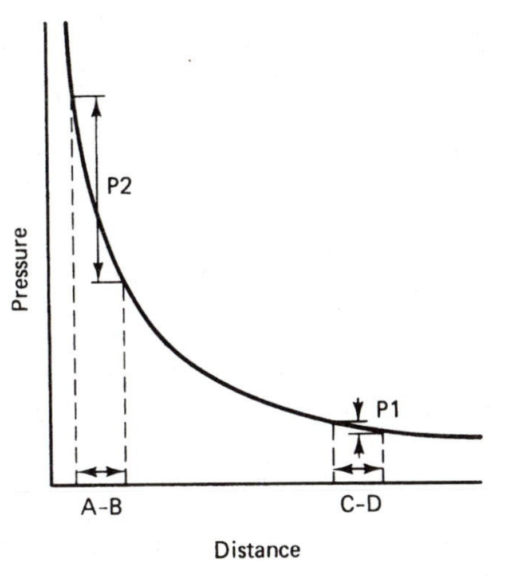

2. Graph based on the inverse square law, showing how rapidly the output from a microphone can rise (P2) with a comparatively small decrease in distance (A–B) when very near the source. This factor and the one above illustrate the vast range in volume and change in quality that can occur with very close working. Compare the small increase (P1) resulting from the same movement (C–D) at a greater distance.

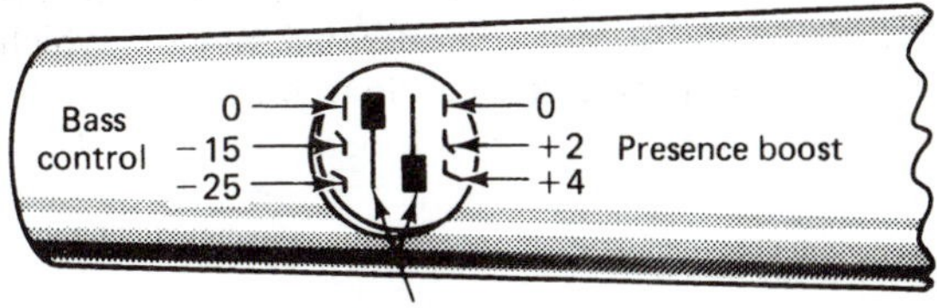

3. Section of microphone casing showing bass control and presence boost switching.

Personal microphones are useful in mobile situations.

Personal, Lavalier or Clip-on Microphones

There are several basic methods by which sound can be picked up from a person who is moving about.

Microphone boom

From the sound quality point of view, mobile sound sources are best dealt with by suspending the microphone on a telescopic boom or on a smaller 'fishing rod'. These can be moved about to follow the artist while maintaining the correct relationship between them.

Hand-held microphones

Hand-held microphones must be robust. They are usually provided with close-talking windshields and built-in filters to reduce the bass boost and possibly high frequency sibilance caused by close working.

Lavalier microphones

Miniature microphones, which need not be held, can be hung around the artist's neck by a lanyard (lavalier) or clipped to the clothing. These microphones can be either crystal, miniature moving-coil or, in the case of the better-quality ones, electrostatic.

The electrostatic microphones generally use electret (permanently polarized) diaphragms to eliminate the need for a polarizing voltage supply. They do, however, need a head amplifier and this is usually arranged in the form of a small cylinder which can be housed in a pocket and connected to the microphone by a short length of thin flexible cable. The connection to the amplifier is usually a much heavier cable which can stand up to being dragged along the floor.

This type of personal microphone can produce quite good results, provided that it is carefully positioned on the person and is not shielded by the clothing. It is, however, an unnatural position from which to pick up the voice; there may be bass boom (due to chest resonance), excessive nasal tone and lack of sibilants as the nose is in line with the microphone which may be shielded from the mouth by the chin. This can give the sound a distant 'off mike' quality. For these reasons, some personal microphones are designed with 'presence' peaks and a diminishing bass response, others have external filters to modify any flat characteristic and, in one type, the response can be adjusted by altering the position of the lanyard clip, which forms a cavity resonance.

Wireless microphones

Both hand-held and lavalier microphones suffer from the disadvantage that they employ a cable which can restrict the artist's movement. Radio transmitters are available of about the size of a cigarette package into which personal microphones can be connected (also eliminating the need for head amplifiers in the electrostatic type); see page 50.

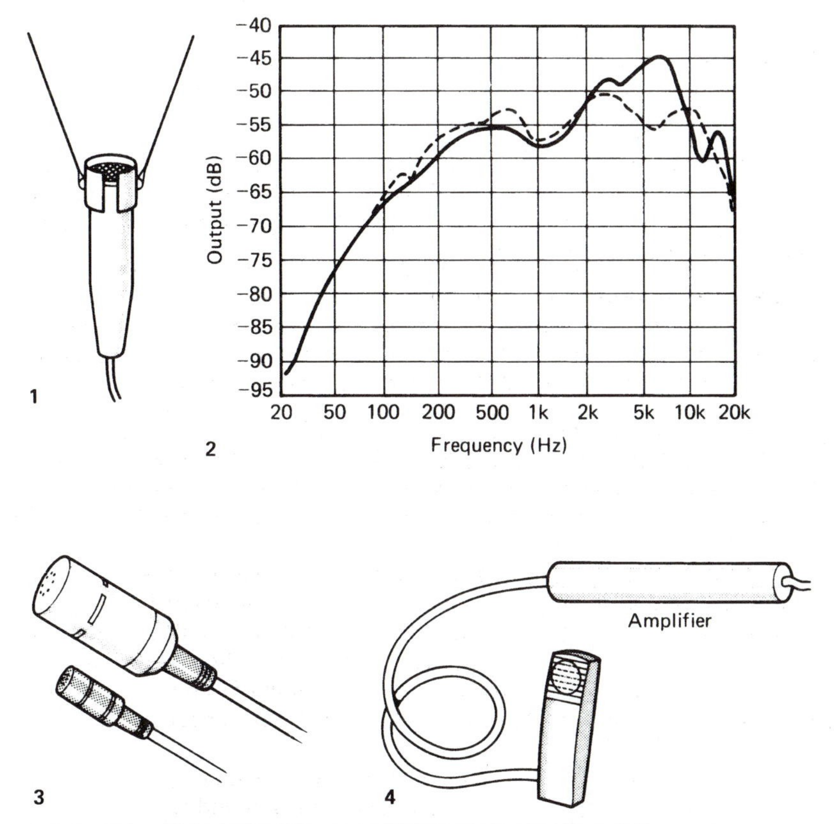

1. A personal lanyard microphone incorporating a movable clip.

2. Illustrating the effect of moving the clip. When the clip extends over the top of the microphone a cavity resonance effect causes an increase in the response at around 6 kHz giving increased 'presence'.

3. A miniature electrostatic personal microphone which uses an electret diaphragm, with a smaller and less conspicuous (though possibly less reliable) version.

4. An electrostatic microphone in which the diaphragm faces away from the artist wearing it. The idea is to give preference to the other person in an interview situation.

Close-Working Microphones

It is a fundamental fact that microphones that work on the pressure gradient principle produce a marked increase in low frequency response when used close to a point source (e.g. a voice). There are two reasons for this:

1. At close range a point source produces a spherical wave front, which increases the pressure gradient effect.
2. Following the inverse square law (i.e. sound pressure is inversely proportional to the square of the distance from the sound source), the change in pressure over the distance between the front and back of the diaphragm increases at close range. These two factors combine to increase the pressure difference.

At normal working distances pressure gradient operation tends to favour high frequencies, because the phase change within a given distance increases with frequency. This is taken into consideration in the design of microphones with a flat response. The increase in response due to close working has much more effect at low frequencies, where the phase change is small and the pressure difference effect relatively greater. As a result the low frequency response increases progressively as the distance from the source is reduced.

Noise-cancelling microphones

The noise-cancelling principle is widely used for commentator's microphones in noisy situations and for musicians who sing and play their instruments (particularly drums) at the same time. As bass accentuation is a feature of pressure gradient operation the more the microphone relies on this element the more effective it will be. So the effect is most pronounced with figure-of-eight configuration, followed by hypercardioid. Some commentator's microphones have ribbon elements, with the ribbon shielded from the mouth by the magnet assembly, to guard against breath puffs and explosive consonants. The microphone is held at a distance of about 6 cm (2¼ in) from the mouth by a guard which is held above the upper lip. A fine wire mesh on top acts as a nasal windscreen.

Headgear microphones

Another type of noise-cancelling microphone consists of a miniature figure-of-eight or hypercardioid element mounted in a small boom arm attached to headgear, which holds it at the correct distance while leaving the hands free.

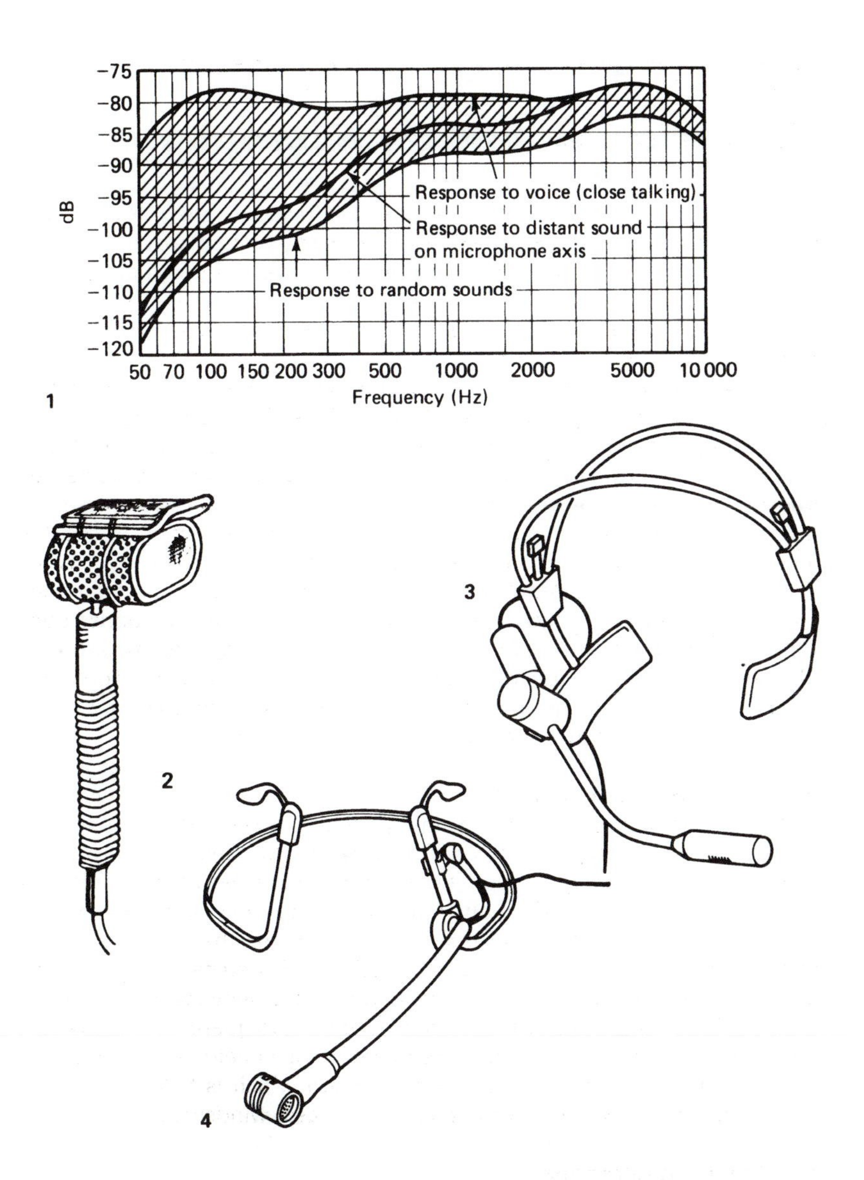

1. Typical frequency response curve (impedance = 30 Ω), illustrating the discrimination between close speech and distant random noise of a pressure gradient microphone. The discrimination is especially good for low frequencies, owing to the bass tip-up effect which affects close working. The equalization provided in the microphone output levels the bass increase for the voice and in so doing makes the microphone less sensitive to background noise.

2. A noise-cancelling ribbon microphone. The bar at the top is held against the upper lip to fix the distance from the mouth. The handle houses a transformer.

3. A noise-cancelling microphone fitted to a headband.

4. A noise-cancelling microphone with a behind-the-neck band which is especially useful for vocalists who play instruments and move about.

Acoustical Boundary Microphones

The acoustical boundary, boundary layer or pressure zone microphone consists of a small microphone element (usually self-polarized electrostatic) mounted close to, or recessed into, a baffle which can be placed on a table or floor or fixed onto any large reflecting surface. The principle of operation exploits the pressure-doubling effect that occurs close to a large sound-reflecting surface, owing to the sound waves effectively doubling back on themselves. If a sound wave encounters a rigid surface which reflects it back along its original path, there can be no movement of the air particles past that point. This is effectively a node (zero point) for particle displacement. But the pressure wave is maximum when particle displacement is zero, so there are pressure wave anti-nodes (maxima) at the boundary surface. Moreover the reflected waves are returned in phase with the incident wave so the resultant pressure is doubled.

All this assumes that (a) the surface is perfectly reflecting, (b) it is rigid (i.e. the incident sound wave does not cause it to vibrate) and (c) the reflective surface is larger than the wavelength of the sound. If these conditions are not completely met the reflected wave is smaller and possibly slightly out of phase with the incident wave so the combined effect is correspondingly smaller.

Some boundary microphones consist of pressure elements recessed just below the surface of a small metal or wood baffle. These have an hemispherical response, i.e. half omnidirectional, with an output sensitivity 6 dB higher than a conventional omnidirectional microphone in the same position. An alternative version employs a small microphone capsule raised just above the surface of the capsule with the choice of half-spherical or half-cardioid response. The latter can be helpful in reducing the risk of PA feedback in some circumstances, such as when used on stage. Some have built-in filters to suppress footfalls and other low-frequency vibrations.

One design problem is that sound waves impinging on the edges of the plate in which the microphone is mounted are diffracted to form an additional spherical sound source. The extent to which the incident and refracted sound waves react together, and the effect on the frequency response, depend upon the frequency and angle of incidence. The effect can be mitigated by chamfering the edges of the microphone plate. It is claimed that an effective solution is to make the shape of the plate and the position of the transducer such that the paths between each part of the edge and the transducer are distributed evenly among a range of lengths – hence the triangular shape of the Neumann condenser boundary-layer microphone and the asymmetrical position of the transducer. This microphone incorporates a small amount of HF lift, intended to compensate for HF losses created by large working distances and the fact that it is likely to be used on a surface (e.g. the floor) to which the sound source is not directed.

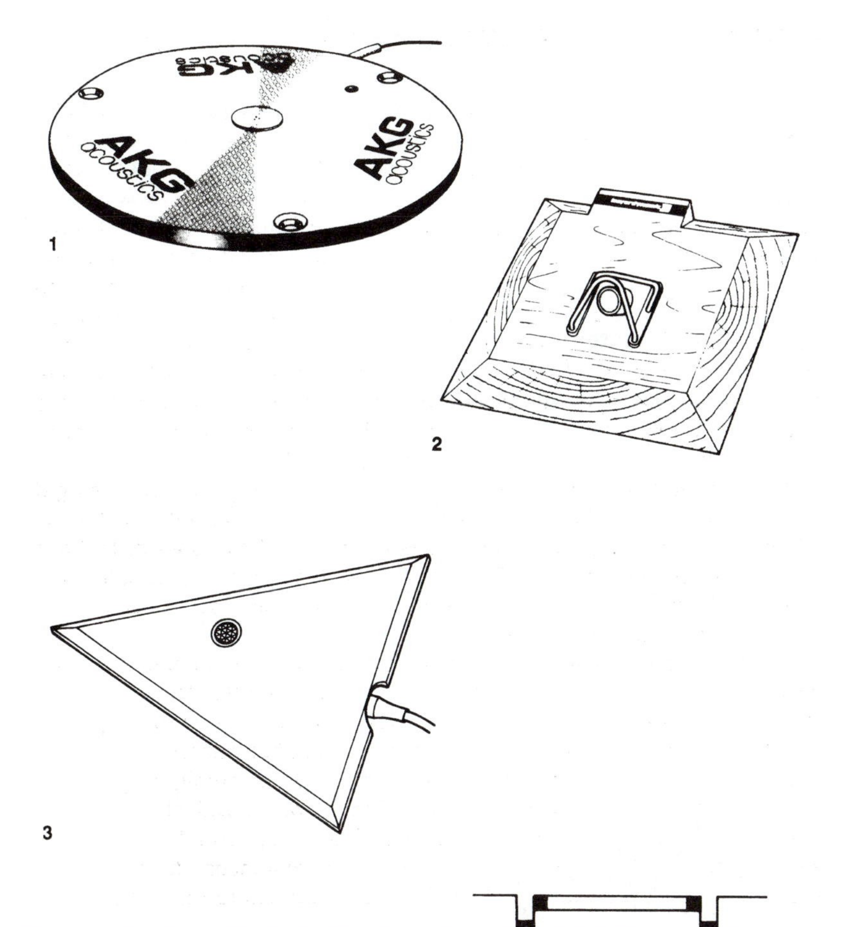

1. One type of acoustical boundary microphone consists of a metal disc 160 mm in diameter and 6 mm thick (6.3 × 0.25 in). A condenser element is completely recessed into the centre of the disc, giving a half-spherical response (Courtesy AKG).

2. Another type has the microphone mounted in a slanting position just above the surface of a 220 mm (8.5 in) square wooden plate, protected from damage by a metal bow. A choice of capsules is provided to give half-spherical or half-cardioid response (Courtesy Bayer).

3. Triangular boundary microphone (Courtesy F. W. O. Bauch).

4. Cross section of a flush-mounted capsule, showing the flexible suspensions designed to reduce the effect of footfalls or mechanical shocks.

Miniature radio transmitters can obviate the need for microphone cables.

Wireless Microphones

When an artist has a large area of movement, or when a cable would be an encumbrance, a wireless microphone can be used. This usually consists of a small transmitter, carried on the person of the artist, connected to a personal microphone (either clip-on or hand-held). The transmitter is usually carried in the pocket or in a pouch tied around the waist. There are also radio microphones in which microphone and transmitter are built into the same case in the form of a baton to be held in the hand.

Transmission systems

The frequency-modulated signal is picked up by a receiver which is normally housed in the sound control room. The choice of receiving aerial depends on the circumstances. If only one artist, with a limited range of movement, is involved, a single dipole aerial can be used, mounted as near as practicable to the action. If several artists with radio microphones are required to move about extensively, it may be advisable to work their respective aerials at some distance (e.g. 10 metres or 30 feet away from the action) and well separated from each other to avoid 'frequency pulling' of the receivers by the other transmitters.

Long range reception (e.g. an operator walking with a shotgun microphone and transmitter to cover sound effects for a golf match) may require a highly directional aerial panned to follow the action. In difficult circumstances, where large areas of movement are involved and there are multiple reflections from moving cameras and other metal structures, a form of 'diversity reception' is required to avoid the effect of 'nodes' (positions of signal cancellation) due to out-of-phase reflections. This can be a simple matter of combining a number of aerials into one receiver, or a system of switching between two aerials so that the aerial providing the best signal is always instantly selected.

The receiver should have a limited AGC or a muting circuit to prevent sudden increases in noise level in the absence of signal, which is much worse than momentary silence. In the ultimate a sophisticated diversity receiver would be used which measures the output from several aerials and automatically switches to the best one from instant to instant or combines their outputs. When a number of radio microphones are to be used in close proximity they can tend to pick up each other's signals and re-radiate them as intermodulation products. So it is important to choose frequencies which do not coincide with the sum or difference frequencies of any of the others (i.e. $2f_1 - f_2$ or $2f_2 - f_1$).

Site surveys

Watch out for large transmitters in the vicinity, police and other radio intercom systems, neon lighting and other sources of electrical interference, such as HF noises due to digital appliances (synthesizers and effects units etc.). Filter units are available which should be connected in the inputs and outputs of such equipment if they cause interference.

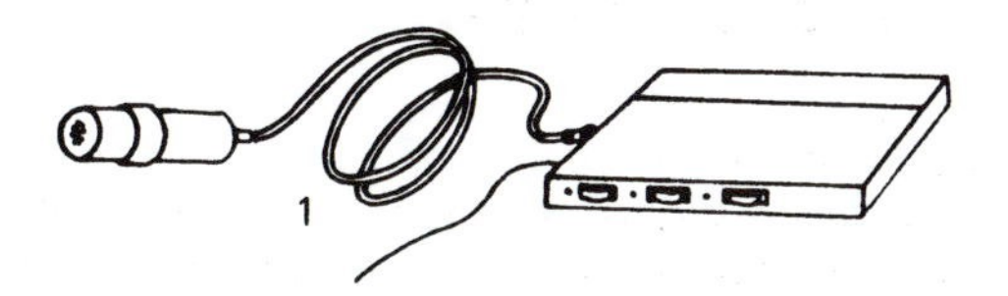

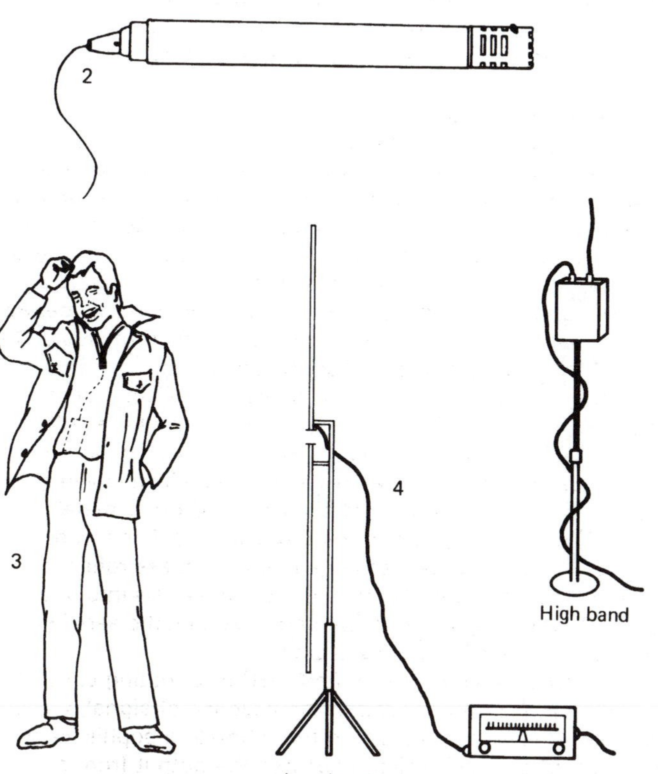

1. Wireless microphone for studio use. The transmitter is miniaturized and made as inconspicuous as possible. The microphone can be any of the personal or hand-held types. The miniature clip-on type is especially suitable.

2. Baton microphone for interviews. This type can be handed from person to person easily.

3. Method of accommodating a wireless microphone.

4. The aerial should be as close as possible to the action. A portable stand can be useful. For theatre work it is often possible to get the aerial close to the artist by positioning it under the stage.

Super-directional Microphones

There are many applications where microphones have to be used at a considerable distance from the sound source. These include radio and television production covering sporting events and nature recording etc. The requirement is for a microphone with a very narrow acceptance angle to give maximum discrimination between the wanted source and the ambient noise and reverberation.

The reflector microphone

The simplest and most efficient method of producing a very narrow angle of acceptance is to place the microphone at the focus of a parabolic reflector. This arrangement is very efficient because the reflector acts like a large scoop which collects the sound and focuses a narrow beam on to the face of the microphone. The best type of microphone to use for this purpose is one with cardioid response arranged so that the live face looks into the reflector, encompassing all of it in its acceptance angle, and the back faces the wanted source of sound. This may seem odd until it is realized that it is a feature of the parabolic shape that the path lengths of all the waves from a source straight ahead to the point of focus are equal, so that they reach the microphone face in phase. Any sound picked up direct would travel a lesser distance and, being out of phase, would tend to cancel rather than augment the reflected sound.

There are two reasons for using as large a reflector as possible:

1. The larger the reflector the more of the sound wave it will collect and scoop into the microphone and thus the greater the forward sensitivity.
2. Surfaces can only reflect sounds of wavelength shorter than their dimensions (see page 14), so that for a parabolic reflector to operate through the full audio range it would appear to need to be of the order of 9 m (30 ft) in diameter!

In practice an acoustic effect comes to our aid. The sound pressure at the centre of a flat circular disc is greater than that in free air because sound waves are reflected at the surface and at the circumference of the disc. The reflected waves add at the centre of the disc and can increase the pressure by up to three times, i.e. up to 10 dB. As the microphone is near the surface most of this gain is available to maintain the directional response down to a much lower frequency than the dimensions would suggest. Practical parabolic reflectors are usually about 1 m (3 ft) in diameter.

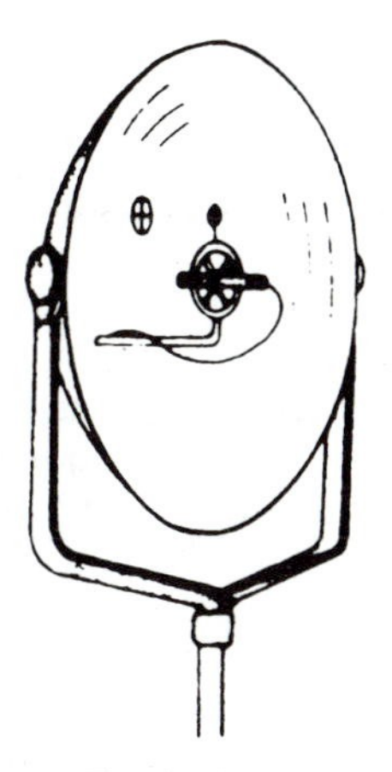

Microphone in parabolic reflector. The cardioid microphone is mounted facing into the dish and away from the source of sound. Note sight hole and cursor for aiming the dish.

The angle of pickup of a parabolic reflector microphone depends on the sharpness of the focus, which also depends on the size of the diaphragm. With a small diameter diaphragm the angle can be so narrow that, unless the wanted sound source is at a considerable distance, it is necessary to defocus the microphone slightly. Parabolic reflectors are usually provided with gunsights to enable them to be aimed accurately.

When using parabolic microphones for sound effects at sporting events it is advisable to mount the dish in a high position, looking down, otherwise it is perfectly possible to pick up crowd conversation from the other side of the stadium.

Even a small parabolic dish can be extremely cumbersome, especially when used out of doors in a high wind.

The requirement for increased microphone working distances has led to the development of the shotgun microphone.

The Shotgun Microphone

There is frequently a requirement for microphones with a sufficiently narrow angle of acceptance to be able to pick up sound from a distance while discriminating against ambient sounds, especially in television, where microphones often have to be used at a distance to keep them out of the picture, and for recording effects and speech in the open air in circumstances where it is not possible to get close to the subject.

The parabolic reflector, described on p. 52, is very efficient as a means of focusing on distant sounds and is often used for nature sound recording. But even a 45 cm (18 in) reflector can be a cumbersome piece of equipment. The search for a more convenient form of narrow-angle microphone has led to the development of the interference tube or 'shotgun' microphone.

Sensitivity

Unlike the parabolic reflector, which collects and concentrates the wanted sound, the shotgun mic works by rejecting unwanted sounds. It is in fact no more sensitive than a conventional microphone using a similar transducer.

Acoustic interference

The shotgun mic consists of a microphone element coupled to the end of an acoustic interference tube. The tube has a large number of slots at right angles along its length. It works on the principle that sound waves arriving in line with its principal axis go straight down the tube to the microphone and are practically unaffected, whereas sound waves arriving from other angles enter through the slots and, according to the distance of each slot from the diaphragm, travel different distances and arrive in differing phase, thereby tending to cancel each other out. Clever design of acoustic damping and matching to the microphone element helps to control standing waves and equalize pressures at the diaphragm.

Acceptance angle

As with all acoustic devices, interference tubes are only completely effective for sounds with wavelengths shorter than their length. At lower frequencies the directional response reverts to that of the microphone to which the tube is attached (usually cardioid or hypercardioid). Most shotgun microphones have condenser (electrostatic) elements because of their high sensitivity and small size, but some are available which have high-output dynamic elements, especially suited to fieldwork where robustness is a prime consideration.

The interference principle is only really effective in free-field conditions and with plane waves (e.g. at a distance in the open air). It is not so effective against reverberation where the sound comes from all directions in random phase.

1

2

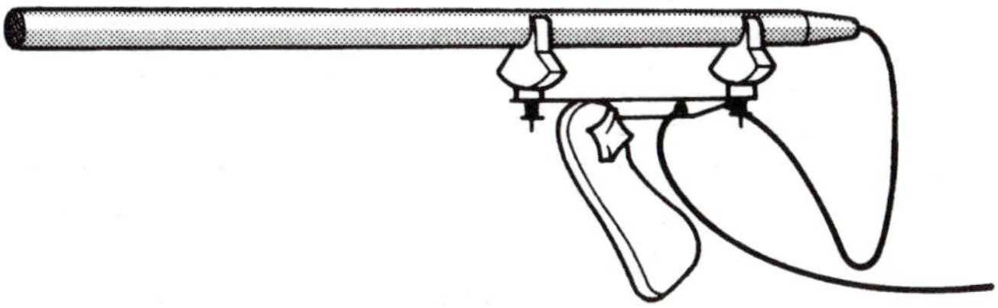

3

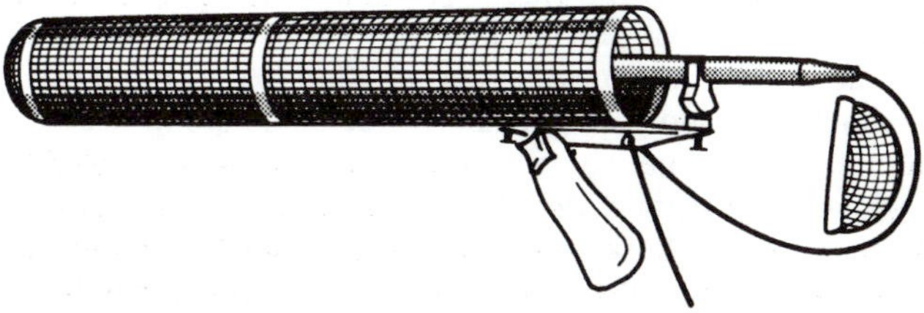

1. Shotgun microphone. Note the sound-entry slots in the tube.
2. Shotgun microphone with pistol grip.
3. Shotgun microphone in windscreen.

Shotgun microphones are available in a variety of sizes from about 173 cm (5 ft 8 in) to 23 cm (9 in). The longer the better as far as directivity is concerned, but this has to be weighed against convenience, taking into account the purpose for which it is to be used. The longest shotgun tends to be cumbersome and its use is mainly restricted to distant effects coverage and audience participation programmes.

Microphones with interference tubes about 40 cm (16 in) long are especially useful in association with vision. They can be hand-held with a pistol grip, or mounted on booms, fishpoles or cameras. They are effective at medium range, having an acceptance angle that varies between about 50° at high frequencies, widening as the frequency decreases until, at about 500 Hz, the tube ceases to be effective and the response reverts to that of the microphone element alone (cardioid or hypercardioid as the case may be).

Microphones with small interference tubes have less directional qualities but can be useful on fishpoles or hand-held for interviews, where they can provide reasonable separation from ambient sound without requiring accurate aim. They are also useful for isolation from PA.

Windscreens

Most microphones are affected by wind. The more directional they are, i.e. the more they rely upon the pressure gradient principle, the more susceptible they are to wind.

It is an unfortunate fact that wind on a microphone does not make the sort of sound normally associated with wind, which might at least explain its presence, but a thumping or low rumbling sound – more like thunder.

There are two major causes of wind noise: (a) rapidly moving air actually hitting the diaphragm and (b) turbulence around sharp contours of the case, causing eddies to occur which penetrate the shielding to reach the diaphragm.

The diaphragm can be protected from the effect of wind blast by screening it with an acoustically porous material such as polyester foam or several layers of fine mesh. Such a windscreen has much less effect on the sound than it does on the wind, because the unilateral velocity of even a light wind of 10 km/h, 2.8 m/s (6 miles/h, 9 ft/s) is much higher than the vibratory velocity of sound waves (less than about 8 mm/s or 0.3 in/s) and the flow resistance of porous material increases with particle velocity. Nevertheless there will be some degradation in the sound quality (usually a reduction in HF sensitivity) when windshields are in use.

The effect of turbulence can be greatly reduced by creating a smooth airflow over the surface, which involves increasing the radius of the curves ideally to a spherical or cylindrical/spherical section. Unfortunately, to be effective against strong winds windscreens must be large in relation to the microphone; that can mean diameters of 10–15 cm (4–6 in), which tends to preclude their use in shot.

Pop windscreens

When speakers work close to the microphone there is a risk of breath puffs, particularly from 'explosive' consonants (P and B) which can produce air particle velocities of over 90 m/s (300 in/s) resulting in very loud thumps in the sound. Some microphones have close-talking shields built in to their construction, usually layers of fine mesh or 'sintered' material. Otherwise small windshields are available made of plastic foam supplied in various colours to match the costumes.

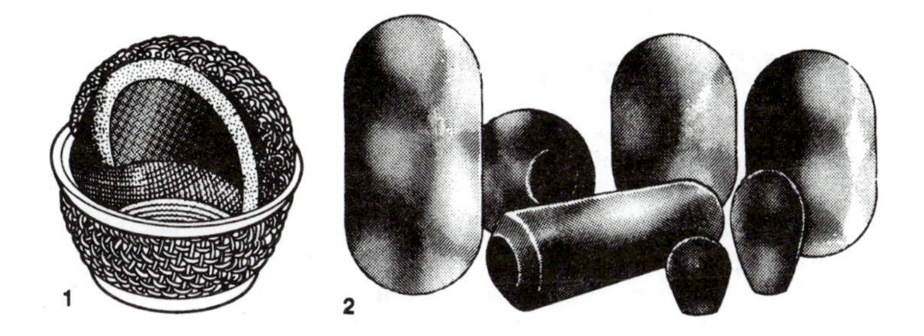

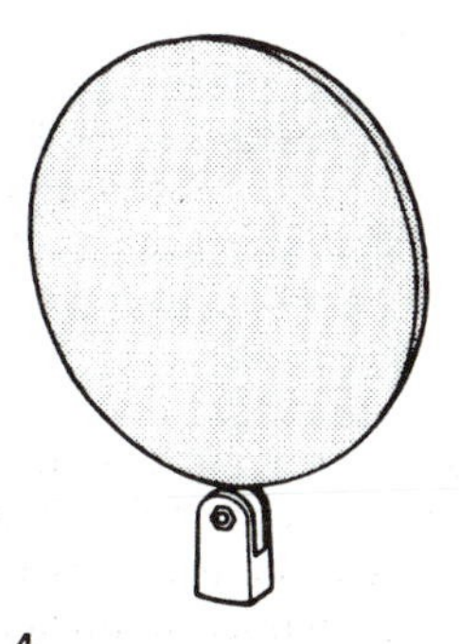

1. Section through an integral windscreen for a vocalist's microphone, which might be subjected to rough usage as well as close working. Two layers of strong wire mesh and one of fine wire mesh are followed by a layer of plastic foam. The microphone element is also resiliently mounted to absorb shocks and handling noise.

2. A selection of plastic foam windscreens to suit different types of microphones. They are available in a variety of colours.

3. Windscreens for gun microphones, showing end cover removed and additional 'windjammer' cover, which can be zipped over the outside to protect against severe wind.

4. A stocking-type pop screen keeps working distances constant.

There are no hard and fast rules about the choice and positioning of microphones but here are some general points.

Using Microphones

How many microphones?

The golden rule about the number of microphones to use is *the fewer the better* because, unless the source for each microphone is isolated to it, the overlapping sound fields will reduce the clarity of the sound. Nevertheless, the fact remains that there are many circumstances where the desired quality of sound can only be achieved by multi-microphone technique. At one end of the scale it is possible, given the right conditions, to record a symphony orchestra of 80 musicians on one stereo pair, while at the other there are 'pop' groups that use 10 microphones on a drum kit. The factors governing the number of microphones are: the internal balance of the source; the closeness of the technique required to produce the desired quality of sound; and whether different elements in the balance require different degrees of processing, i.e. frequency response shaping, reverberation, etc.

Internal balance

Very few programme sources are sufficiently balanced within themselves for recording purposes, bearing in mind the limitations of reproduction in the domestic situation (see p. 62). Even in the case of the symphony orchestra, mentioned above, the conductor would expect to receive guidance on points of musical balance from the sound balancer because, standing so close to the orchestra, it is not possible to assess the balance between the musicians with sufficient accuracy for recording purposes. The majority of musical combinations are not at all well balanced and require each section or musician to have a separate microphone which can be separately controlled to obtain a balanced mixture at each moment in time. This multi-microphone technique throws considerable responsibility on the balancer but in the professional recording business they tend to be musicians, specially trained for the purpose.

Sound character

The character of the sound depends upon the type of microphone, its proximity to the source and the technical processing applied to its output. The main factors governing closeness are the ratio of incident to ambient sound, i.e. separation from reverberation and other sources, and sound quality. Most instruments produce differing frequency characteristics from different parts and in different directions. At varying distances these frequencies suffer differing attenuation and phase relationships. To obtain the sort of crisp attack that is demanded of modern music (reverberation can be added artificially later) and achieve good separation between the instruments a close technique is necessary. This in turn demands multi-microphone technique because it is not possible to get close enough to a number of sources simultaneously.

1. A hypercardioid microphone to give maximum discrimination against PA loudspeakers.

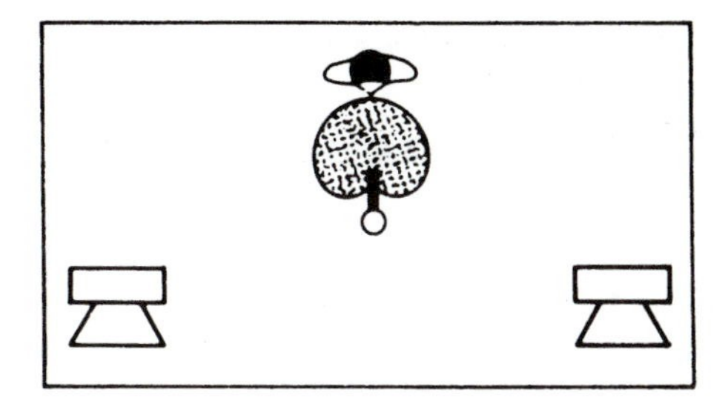

2. A cardioid microphone for a vocalist angled so as to discriminate against the orchestral accompaniment. The angle of discrimination of a cardioid microphone is much more acute at the back than at the front. The small 'fold-back' loudspeaker may be necessary to enable the vocalist to hear himself.

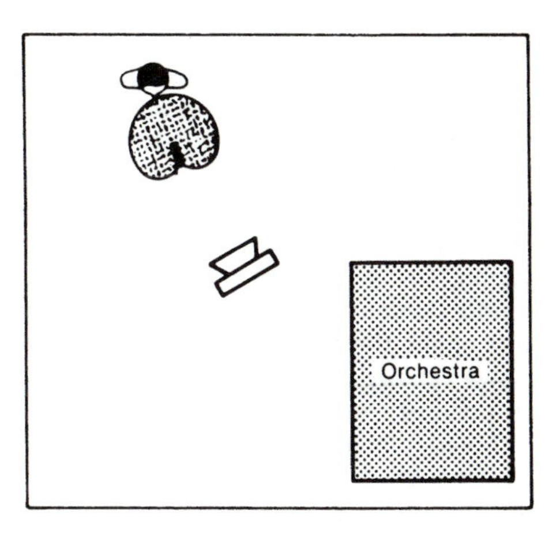

3. The figure-of-eight characteristics of a ribbon microphone used to isolate the comparatively weak sound of a flute against the powerful tones of a trumpet.

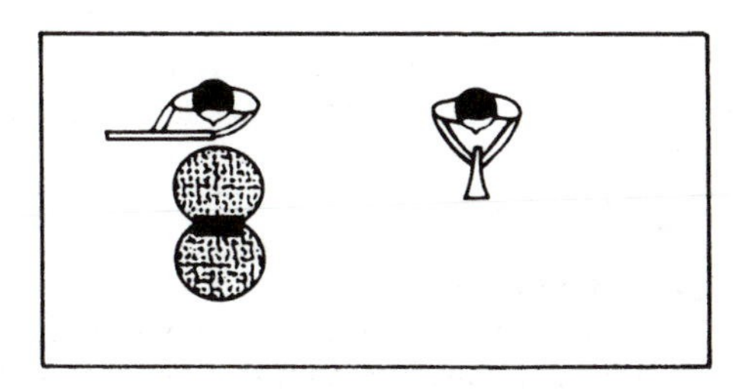

In selecting the type and position of a microphone for a particular purpose it should be remembered that it is often not so important to point the live side of the microphone directly towards the wanted source as the dead side towards the source to be discriminated against.

When using more than one microphone it is important that they are properly phased. This can be checked by placing them near to each other and speaking between them. If they are out of phase the sound will become distorted and weaker when they are both faded up.

Sound Shaping

No matter how good the audio equipment in use, there will be many occasions when the sound can be improved by using equalization and various 'sound sweetening' devices, either to tackle the problems caused by microphone technique imposed by vision considerations or for artistic effect.

Equalization

Most sound control desks incorporate equalizers for shaping the frequency response of individual channels. These can be used to equalize (as the name suggests) the output to produce a flat response, or to enhance some aspect of it.

In considering the use of EQ it is worth remembering the characteristics of our hearing. Our ears are most sensitive to frequencies in the 3–5 kHz region, so an increase in this area tends to give greater *impact* to the sound, i.e. more of an edge to cut through other sounds. Also this is the area where the sibilants in speech and vocal sounds are most prominent. It gives clarity of diction and a sense of *presence* or closeness. This is the purpose of the so-called *presence filters* (usually parametric equalizers with variable width (Q) and height peaks that can be moved up or down the frequency scale to suit the voice). In general we tend to associate closeness with the ability to hear the high frequencies because these are the more attenuated with distance by friction with the air. On the other hand too much HF boost, particularly in the 3–5 kHz region, can give the sound an unpleasant harsh quality which soon becomes tiresome to the listener. Bass boost can give the sound added depth and sonority, but too much can make it dull and put a strain on the reproducing ability of the average domestic receiver, resulting in some unpleasant noises.

Equalizers

These are necessary to correct for deficiencies in audio response. For example, a clip-on microphone used under clothing could be lacking in HF and 'presence'. The output could be compared with a regular mic in a good position, say in a boom, and the response adjusted with a *graphic equalizer*, which gives separate frequency response control of each octave, to match the sound.

Other correcting equalizers include *shelving filters*, which provide a flat response for most of the range with a rapid cut-off at the top or bottom, useful for tackling spurious noises like ventilation noise and for producing telephone and intercom effects. *Notch filters* provide very steep dips in the response at specific frequencies and are useful for removing mains hum or resonances.

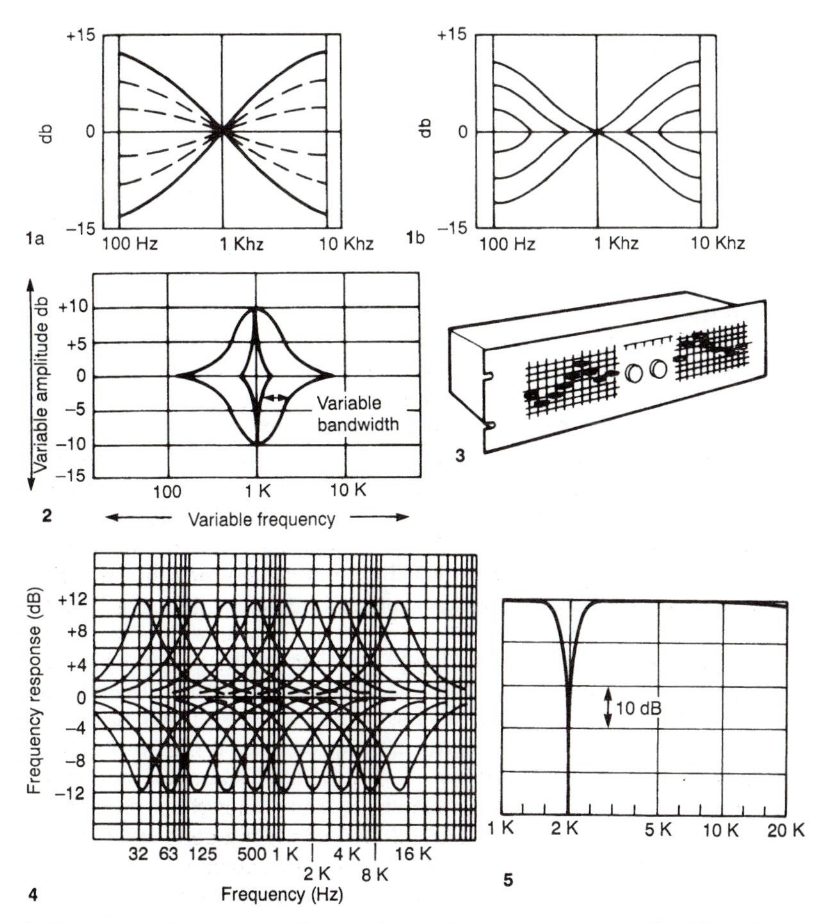

1. *Response selection amplifiers.* (a) In one type the bass and treble lift/cut is adjusted by varying the slope of the response curves about a fixed middle point, in this case 1 kHz. (b) In another type the bass and treble lift/cut is altered by varying the turnover frequencies.

2. *Parametric equalizers.* The bandwidth of the peak or trough can be adjusted between about 0.1 and 5 octaves. Its amplitude can be adjusted between zero and about ±15 dB and its position slid along the frequency scale. The width of the peak or trough (Q factor) can also be adjusted.

3. *A typical stereo graphic equalizer.* This type of equalizer gives control to ±12 dB over 10 octaves centred on 32, 63, 125, 250, 500, 1 k, 2 k, 4 k, 8 k and 16 kHz. The positions of the sliders give a 'graphic' indication of the shape of the response curve.

4. *Graphic equalizer curves.* Careful design is necessary to prevent phasing distortion due to interaction between the overlapping peaks. Most graphic equalizers produce a response that looks very distorted when viewed on an oscilloscope but with good design this is not audible.

5. Notch filter.

Sound Control

Most reproducing equipment has a volume control, so neither the artist nor the sound operator can have any say over the actual volume at which their material is finally reproduced. They can merely affect the *relative* loudness of the various parts of the programme and its individual elements.

Basic requirements

There are three basic requirements:

1. To keep within the dynamic range of the system. In most cases, such as recording and live transmission, the upper limit is set by the onset of overload distortion and the lower by the threshold of intrusive noise.
2. To adjust the scale of volume range to suit the listening conditions and environment. Most types of programme material, from music to drama, involve a range of volume that would be quite unacceptable in average domestic circumstances.
3. To ensure that items produced in different areas and at different times match each other and in broadcasting maintain a smooth balance between successive programmes.

Thus there are two aspects of volume – the technical and the artistic. As we are not able to measure absolute sound levels with our ears and have a poor memory even for relative levels, we need a meter (as described on page 66) and a set of rules regarding the proper level for each type of programme material if we are to keep within the technical limits and provide a balance between successive items or programmes. But the meter does not help us to make artistic judgements or even assess loudness, which is largely subjective.

Use of loudspeakers

The only way to effective programme control is by means of a good quality loudspeaker, making use of a meter to 'calibrate' the ear in the way that a car driver judges his speed without watching the speedometer all the time.

The importance of listening levels

Practically all operators professionally engaged in sound control listen at very high volume. This is necessary to achieve the right degree of concentration and because otherwise there is a tendency to 'wind it up' out of sheer enthusiasm. Unfortunately this can give a very unreal picture of what the listener will hear and can lead to a misleading balance, particularly as between vocal and accompaniment or speech and sound effects. It is therefore advisable to 'dim' the loudspeaker from time to time (perhaps occasionally listening on a small loudspeaker of the type fitted in the average television receiver) and get used to making allowances for the difference.

1. *Speech modulation.* Most sources of sound are continually varying and interrupted in character and a fade which continues through the gaps in the material would result in a succession of steps in the output.

2. *Unobtrusive control.* If the intention is for the operation to be as inconspicuous as possible, which is the case for most types of material and particularly music, the action should occur in the gaps in the material or during rapid fluctuation of level.

3. *Control for dramatic effect.* If it is intended to make an obvious smooth fade for dramatic effect (e.g. a slow fade on dialogue is the conventional method of suggesting a change of scene in radio drama) movement of the control must occur only during the material, preferably during sustained phrases.

The Points of Control

Sound control desks come in a variety of sizes and degrees of complexity, to handle any number of channels from less than six to more than forty. Most desks provide several points in the programme chain at which control can be exercised. The method of control depends upon whether the output is to be a complete mono or stereo source or a multi-track recording for subsequent mix-down.

Pre-set channel gain

The range of sound applied to the channel inputs can be enormous, varying perhaps from whispered dialogue to a musical instrument at close range. This can represent a range of up to 60 dB (i.e. 1 000 000:1). To cope with this range the input channels are usually provided with variable attenuators which can be pre-set to adjust the input to the first amplifier to a manageable level. Switchable attenuators are also provided in some microphones to cope with very loud sound sources.

The channel fader

The most important controls are the channel faders, as the whole essence of effective programme control is that it affects only those elements of the sound that require adjustment to maintain balance. Controlling later in the chain could result in the whole level of the balance being varied to accommodate one or two 'rogue' sources. For this reason compression amplifiers (see page 72), if required, should be connected in the channels, or at least the groups, and not in the overall output. Pan-pots are provided at this stage to pan each item into its appointed place in the stereo image.

Group faders

Despite the earlier remarks, there are circumstances where it is logical to group a number of sources into one control – for instance a large chorus, section of an orchestra or even the whole orchestra if it/they have to be balanced against other sources without disturbing their internal balance. This form of control is achieved in some cases by routing the ouput of a number of faders through a designated group fader; alternatively, in 'assignable' desks, channel faders can be linked to operate together, but without losing the ability to 'trim' the mixture with the individual faders.

Master fader

Last in the chain is the master fader (or faders if several outputs are required). This is essentially a back-stop, with which it is possible to set the overall level and fade in and out without affecting the mix.

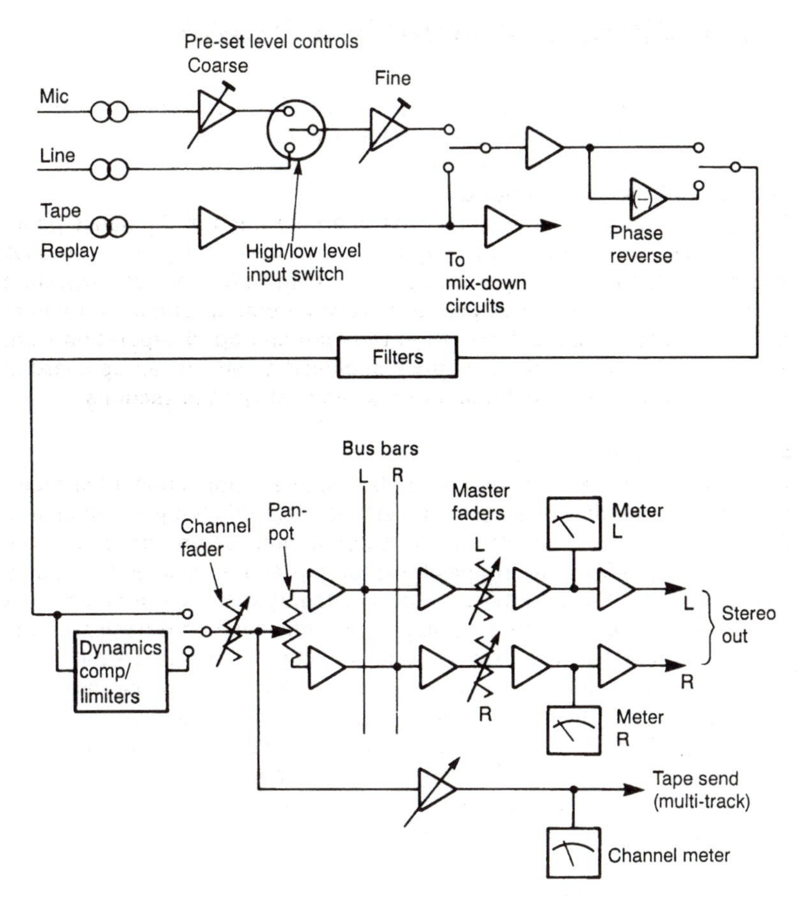

Simplified schematic of a typical sound control desk, showing the points in the programme chain where volume control can be exercised.

In the case of multi-track recording (see page 116) the primary objective is to provide each of the recording tracks with the maximum level of signal, to ensure a good signal/noise ratio, without running into overload distortion (clipping level).

The output of each channel or group is fed to the appropriate track of a multi-track recorder. Meters are provided on the recording channels and in the larger desks on each of the mixer channels for this purpose.

Some desks are divided into two sections, or provision is made to enable a 'dummy' mix with a proper balance to be made while the tracks are being recorded. This helps to check the positioning of the microphones.

Finally, the multi-tracks are played back to the mixer to be mixed into a stereo output, with the same constraints as in a direct recording.

Volume metering is essential for good recording to maintain levels between the parameters system noise and overload.

Volume Indicators

Twin-lamp systems

Some simple recorders employ two lamps (usually light emitting diodes). One lights up to show that minimum recording level has been achieved and the other when the overload point is reached.

Volume unit meter

The majority of tape recorders have volume unit meters to indicate volume. In the simplest form, these can consist of a simple moving-coil meter and rectifier which gives an indication of signal level the accuracy of which varies with the type of programme being measured. All meters can be made to give an accurate indication of zero level (0 VU) on steady tone but the way they react to programme material depends upon the ballistics of the meter (rise time and overshoot) and the value of the resistance through which it is connected. Meters used for professional purposes, where levels have to be set and compared, should conform to the American Standards Association specification. Even these will give widely different, but matching, readings with different material. Due to the slow rise time of the meter, impulsive sounds, such as speech, starting from a low level will read lower than their actual volume level; sustained sounds, i.e. long notes, will read higher and fluctuations near the top of the scale can overshoot and read high.

For this reason VU meters have to be interpreted from experience rather than just read; line up level is usually taken to be +4 dBU (1.73 volts) to read 0 VU to give an actual peak recording level of +8 dBU on average material.

The peak programme meter

A much more accurate visual indication of volume level is the peak programme meter. The PPM has a rapid rise time (2.5 ms) and a slow recovery time (approx 3 s) actuated by charging a capacitor and discharging it through a high resistance.

This action makes the meter easy to read and it gives a reasonably accurate indication of peaks, which is the most important consideration.

Visual displays

The advent of multi-track recording has brought the need for large numbers of meters to be displayed simultaneously. The most convenient way of arranging this is to have indicators consisting of a stack of LEDs which light up in sequence in relation to volume level. Alternatively a plasma, fluorescent bar graph display can be used, or even superimposed on a television picture.

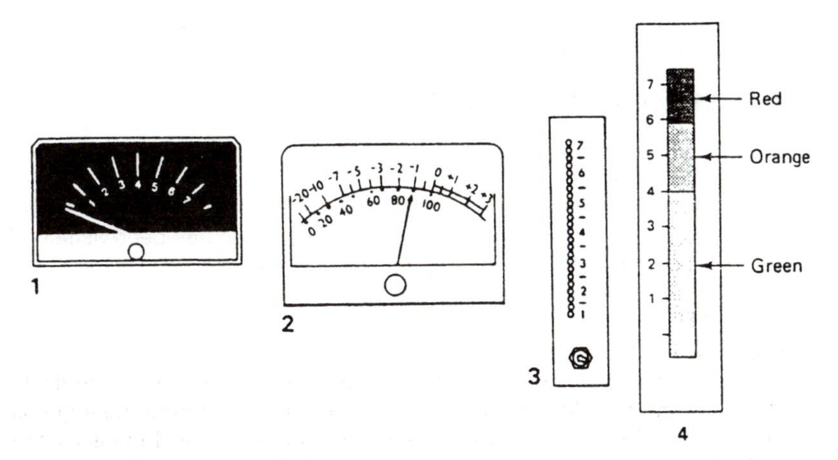

1. *Peak programme meter.* The figures are white on black to reduce eye strain. The law is quasi-logarithmic so that the scale is reasonably evenly spaced, with 4 dB between each of figures 2–7. Line-up level is usually set to 4 and maximum level (+8 dBm) is 6. The extra 4 dB to 7 is to give an indication of the degree of overmodulation.

2. *VU meter.* There are two scales. The upper one is marked in decibels (red above the zero mark, which represents full modulation). The lower scale is marked 0–100 and represents the percentage of full modulation. The scale is uneven, being largely cramped towards the top end.

3. *Visual display meter* suitable for use in individual channels where space is limited. This can consist of a stack of light-emitting diodes. Different colours can be used to represent low, normal and overload levels.

4. *Plasma bar graph.* Various colours are used to indicate low, normal and high levels. This type of meter can usually be switched to display PPM or VU readings or to freeze peak values.

Digital programme meters

Digital meters can work directly with digital mixing consoles or in analogue systems by incorporating a simple DAC (see p. 198). The level is generally displayed as a bar LED or fluorescent indicator. It is normal to interconnect all the individual LEDs in a bar display so that the most significant LED causes all those below it to illuminate also; the display then appears as a bar of light whose length varies with the audio level. The attack time can be virtually instantaneous and the recovery time made to simulate the PPM slow or VU rapid decay or both. In a combined VU/PPM digital level meter the PPM readout is a single illuminated segment which slowly shifts down the scale. On the same display a VU level is displayed as a bar of segments illuminated from the bottom upwards. On tone the display appears as on an ordinary LED meter, but following a transient the length of the bar contracts rapidly leaving a single segment behind to indicate the value of the peak.

Sound Mixing Desks

Sound mixing desks are available in a variety of sizes, from desks with just a few channels that can be used 'in the field' to major studio desks with upwards of forty channels and provision for multi-tracking and subsequent mix-down.

Input arrangements
Most desks are able to accept either low-level inputs from microphones, or high (about zero level) from tape recorders or other pre-recorded sources. This is achieved either by having alternative connection points, with a fixed attenuator pad in the high-level connection, or by pre-set variable attenuators. Provision is usually made to supply +48 V DC in the form of a 'phantom' down the microphone lines to polarize condenser mics or supply direct injection boxes, for taking direct feeds from electronic instruments.

Comprehensive facilities
The larger desks, which incorporate comprehensive facilities, should have cueing systems whereby various items from the input can be 'folded-back' to specific monitor loudspeakers or headphone points in the studio. This can include selective talk-back. Provision is required to supply a large number of outputs from the individual channels (frequently 24, often in excess of 48) for multi-track recording or a mixed-down stereo output.

All channels should have pan-pots, filters and provision for the insertion of dynamics (e.g. compressors) and several sources of reverb. There should be provision to insert graphic equalizers at a later point.

Complete flexibility in routing sources to channels and channels to groups is a considerable advantage.

Assignable desks
By employing voltage-controlled amplifiers (VCAs), which have their gain controlled by a DC voltage, and solid-state switches it is possible to install the control equipment remotely from the sound desk and to arrange that its assignment can be selected according to requirement. This means that instead of individual items being permanently tied to a particular circuit they can be assigned to a variety of functions by remote control. This can lead to great flexibility and efficient use of equipment, reduction in the size of control desks and shorter wiring with fewer connections and less risk of phase displacement. Remote control is especially easy in the case of *digital desks*, where filters can be changed by changing co-efficients. Gain in digital amplifiers is achieved by multiplying each sample value by a fixed coefficient; if it is greater than 1 there is gain, if less a loss (see p. 198). Desk settings can be memorized on floppy disk or multitrack and reproduced for later sessions.

A wide variety of alternative sound consoles is now available, either digital or VCA, offering a similar range of functions at similar cost.

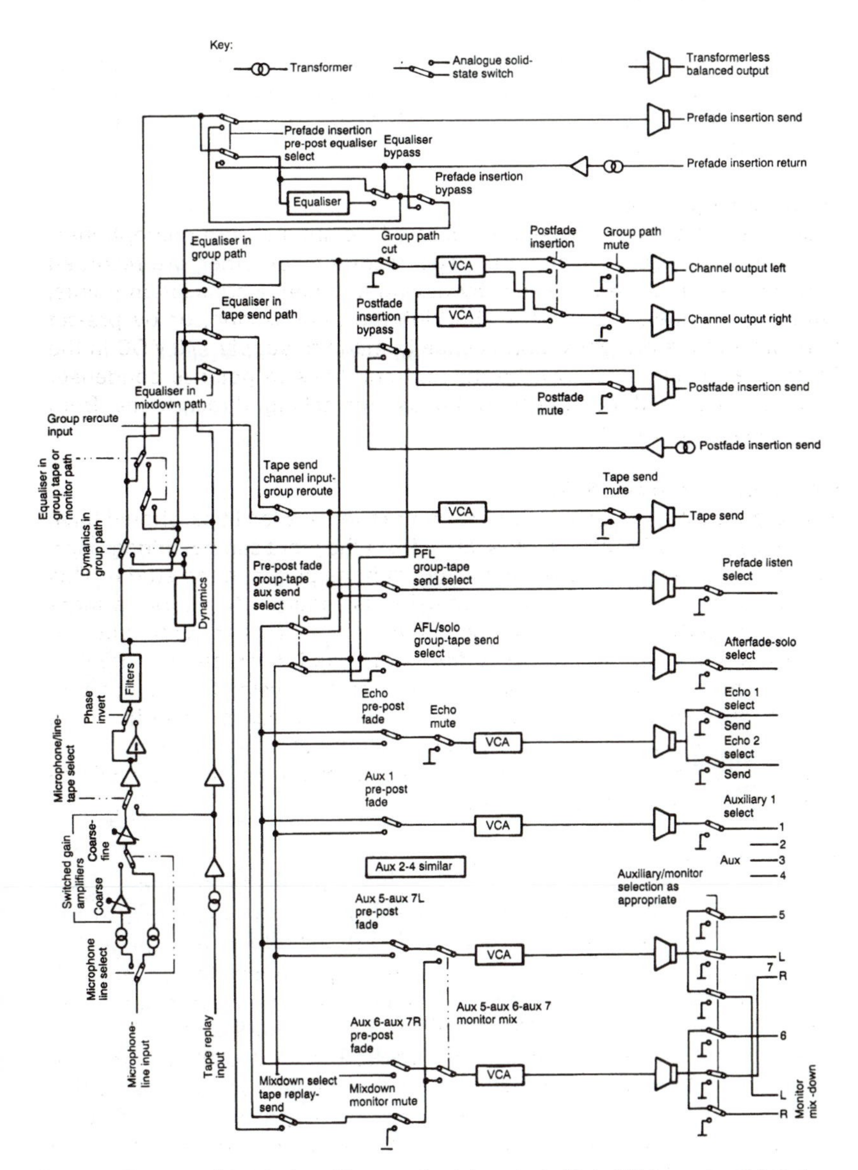

System diagram of an assignable sound mixing desk. Note VCAs and solid-state switches.

Automated Sound Mixing

Where complex music sound mixing is involved and the programme is not live, there is much to be gained by pre-recording, especially if multi-tracking can be used.

Multi-tracking technique

The complexity of modern balance technique is such that it is almost impossible to obtain the best results by mixing the sounds as they are made. It is much better to record each element on separate tracks of a multi-track recorder so that each can be individually treated and, if necessary, repeated without affecting the rest. Moreover, most of the treatment and manipulation of the sound can go on after the musicians have left the studio. However, this 'mix down' process can be tedious and needs to be efficiently handled if optimum results are to be obtained without extravagant use of studio time. This has led to the introduction of *automated mix-down.*

Automated mix-down

During each attempt at mix-down, information regarding the position of each of the faders is stored on tape or in RAM memory in the console, referenced against an SMPTE timecode track stored on the multi-track recorder. This is updated on each successive attempt and when the mix is satisfactorily completed, transferred to be stored off-line on floppy or other disk-based storage.

Mechanical fader control

The method by which the fader settings are reproduced can be mechanical or electronic. In the mechanical system the faders are physically moved by servo motors and can be seen to be operating as though by an invisible human hand. Touching the fader knob has the effect of disabling the fader servo so that the settings can be modified and re-memorized by this natural and instinctive operation.

Electronic channel control

In an alternative method the *effect* of the faders, not their position, is automated. This technique is facilitated by the use of voltage-controlled amplifiers (VCA). These amplifiers have their gain controlled by a DC voltage, typically in the range −2 V to +10 V DC. The control voltage adjusts the amplifier gain logarithmically and is usually arranged so that an increase of one volt represents an attenuation of 10 dB. VCAs are usually set up so that zero volts from the control fader produces zero gain, +10 V gives 100 dB attenuation (cut off) and −2 V gives 20 dB gain. This allows for 20 dB 'headroom' above normal settings.

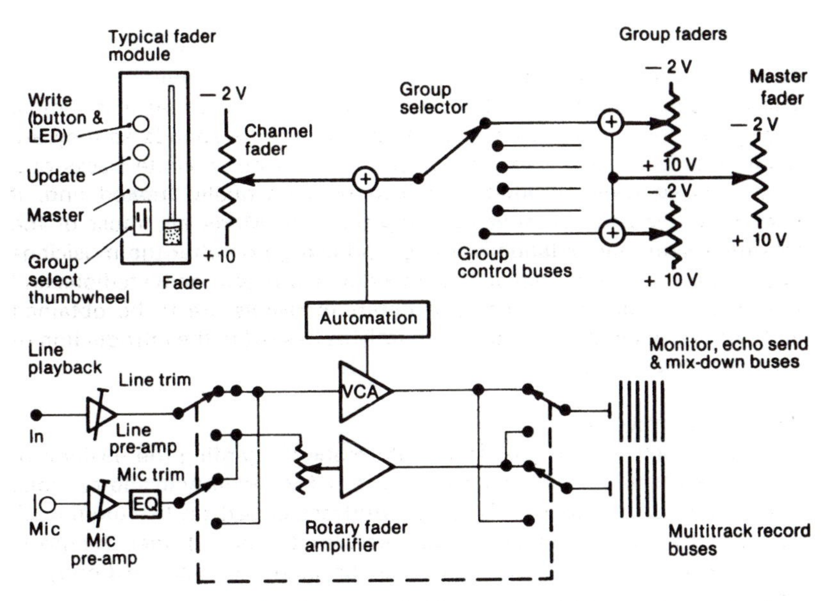

The control voltage for the VCAs is varied by the channel faders. It can also be supplied from group faders by selecting the appropriate bus bar to add to the fader control voltage. Finally the group faders can be controlled by master faders by simply adding their voltage to the groups, the actual volume adjustment still taking place in the channel faders.

One advantage of this system is that the individual channels can be controlled as groups without their outputs being mixed so that they can be recorded completely separately on the multi-track tape.

To make changes in the mix on subsequent attempts it is necessary to find the position on the fader that matches the VCA control voltage. This is shown by two LEDs, one of which lights up when the fader is too high and the other when it is too low. When both light, the fader setting is matched; the 'write' button can be pressed and the fader moved to a new position to update the memory.

Some mixing consoles have alternative paths (controlled by rotary faders) for making the original recording.

Compressors enable higher average sound levels to be maintained without overload.

Automatic Control

No automatic device can provide the necessary aesthetic judgement and finesse for effective programme sound control. Nevertheless automatic control systems (known as compressors and limiters) have considerable application in sound operation practice to assist in the 'taming' of wildly varying and unpredictable sources and to protect equipment from possible overmodulation.

The limiter
The limiter is a device through which programmes can be passed without alteration of the signal until a critical value is reached. If the input signal rises above this value (usually called the onset point), the gain of the system is automatically reduced (below unity) so that the output cannot rise significantly above the limiting value. This limiting action is caused by reduction of the amplifier gain not merely by cutting off the peaks of the waveform (peak-chopping results in very severe distortion).

The compressor
The compressor is similar to the limiter, in that above the onset point the gain of the system is reduced, but its action is less dramatic so that an increase in input level above this value produces a reduced increase in the output. The extent of the gain reduction is usually adjustable and is called the compression ratio. Other controls usually available on compressors are:

1. The onset or threshold point (the value at which the action commences).
2. The attack time, i.e. the time required for the action to take effect following a sudden increase in the level.
3. The recovery time, which is approximately the time required for the gain to be restored to unity after the high level signal has been removed.

The attack time is set as a compromise between too fast, which can alter the shape of the waveform and thereby cause distortion, and too slow with consequent overshoot.

Recovery time is usually set to make the variations of gain as inconspicuous as possible without allowing isolated peaks to depress the level for too long.

The noise gate
A noise gate is almost the opposite of a compressor. It reduces the amplification below a certain threshold level (thereby reducing noise in the absence of signal) and increases it when the minimum acceptable programme level is reached. It thus acts as an 'expander' to increase the distinction between signal and noise level.

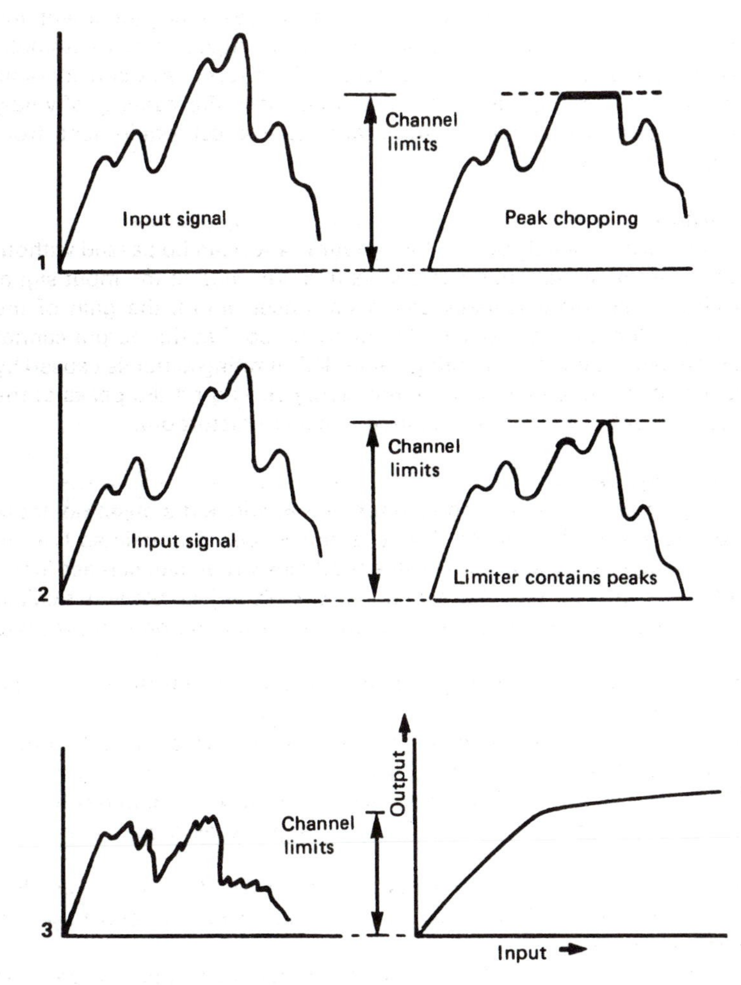

1. Effect of peak chopping with severe distortion.

2. A limiter should cause gain reduction to contain the peak of the waveform.

3. Input/output curve for a limiter showing the 'knee' or onset point of compression.

The relationship between precedence and volume determines the apparent direction of a sound source.

Time Delay: the Haas Effect

There is a relationship between the order of precedence and the relative volume in which sounds are heard that determines the direction from which they appear to come. This is an important consideration in all multi-source sound systems whether they concern stereophony, surround sound or even a simple PA artist reinforcement system.

The Haas effect

Haas investigated the effect on the listener of two sources radiating the same sound from two different directions in the horizontal plane. These could be, for example, two loudspeakers fed with the same signal or a man and a PA loudspeaker reproducing his voice.

If the loudspeaker is nearer to the listener than the man (as often happens with PA systems), the sound from the loudspeaker will arrive first. If the sounds from the two sources have the same volume at the listener's position, his impression will be that all the sound originates only from the loudspeaker and none from the man.

If a time delay of between about 5 and 50 ms is introduced into the feed to the loudspeaker, the impression will be reversed: none of the sound will appear to come from the loudspeaker although this contributes to the volume and reverberation of the sound. If, however, the volume of the delayed sound from the loudspeaker is increased relative to the direct sound from the man, a point will be reached when it overrides the time delay and the sound again appears to come only from the loudspeaker. The relationship between volume level and time delay in establishing the apparent direction of sound is illustrated by the graph opposite.

It will be seen that the maximum effect occurs with delays between about 10 and 30 ms. After about 60 ms transient sounds begin to be discernible as separate echoes. It should be remembered that the speed of sound in air introduces a natural delay of the order of 3 ms/m (1 ms/ft).

Artificial time delay

The ability to insert time delay has enormous advantages, apart from the obvious one in the PA example quoted above. Time delay inserted in artificial reverberation systems improves realism, as it is normal for the first reflections to arrive about 50–100 ms after the direct sound. The ability to make sound appear to come from one direction when the bulk of it is coming from another has considerable artistic possibilities. Time delays can now be provided by digital means.

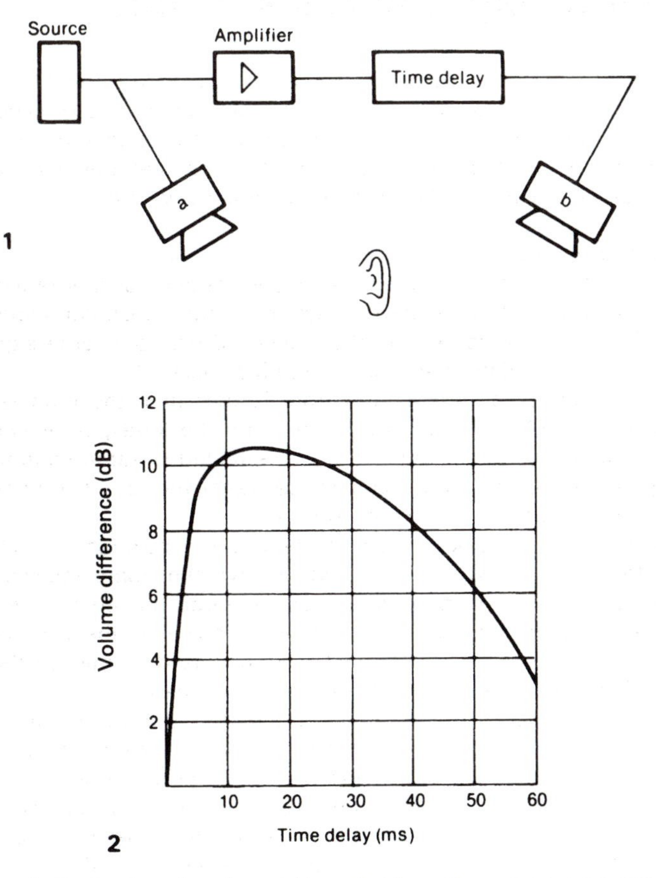

1. If two similar loudspeakers (*a* and *b*) are fed from the same source, one through an additional amplifier and variable time delay, and listened to from an equal distance the following effects will be noticed: (1) If the loudspeakers produce the same volume, in phase, the sound will appear to come from a point equidistant between them. (2) If *b* is made louder than *a* all the sound will appear to come from *b*. (3) If the output of *b* is delayed with respect to *a* but the two volumes are the same, all the sound will appear to come from *a*. (4) The effect of the time delay can be overcome by increasing the amplification to make *b* louder than *a*.

2. The curve shows the relationship between volume and time delay in establishing the apparent position of a sound source. This shows, in the example of 1 above, that if 10 ms delay is inserted in the feed to loudspeaker *b* it would have to be over 10 dB louder than *a* to make the sound appear to come from a point between them.

Receiving separate information in each ear enables us to locate the direction of a source of sound.

Binaural Hearing

So far we have only considered monophonic sound, as though listening with only one ear. Most of us, however, are able to listen with two ears and this gives us the ability to judge the direction of a source of sound and, to a certain extent, to discriminate between sounds coming from different directions.

When a sound comes from a direction other than straight ahead it reaches the nearer ear slightly before the other one. The brain is able to detect this tiny difference in time and also slight difference in intensity and quality (due to the waves being defracted around the head). It can interpret these to gain information about the direction of the sound source. We can also tell whether the sound is coming from in front or behind by virtue of the slight difference in quality that the shape of our ear lobes imparts to sound from each direction. Discrimination in the vertical plane can also be achieved to a limited extent due to the ear lobes and the fact that we seldom hold our heads quite still and exactly vertical.

Aural discrimination

The ability to locate the direction of the various sound sources can help us to concentrate our attention on wanted sounds and subjugate unwanted sounds such as noise or reverberation, etc. If this ability can be applied to sound recording, it can greatly enhance the realism and clarity of the reproduction.

Binaural reproduction

The nearest approach that can be made to the binaural experience by electrical means involves placing two microphones, one in each side of a dummy head which is designed to reproduce as closely as possible the acoustic proportions of the human head.

The outputs of these two channels must then be kept separate (though equal in phase relationship) throughout the entire process of recording and/or transmission and reception to end up each in the appropriate earpiece of a pair of headphones.

The resultant effect can be very realistic, provided that the above 'purist' microphone technique can be employed. It is much less suited to the method of multi-microphone balancing adopted for most stereo recordings. A sound source 'pan-potted' electrically into the middle (see p. 84) can appear to emanate from the middle of one's head in a rather disturbing manner.

A schematic illustration of binaural sound reproduction

A pair of microphones is mounted inside a dummy head, usually made of polystyrene, to copy the appearance of a human head as far as possible. Alternatively, the pair of microphones can be mounted at an angle (usually 90–180°) with a perspex baffle of approximately 30 cm diameter between them.

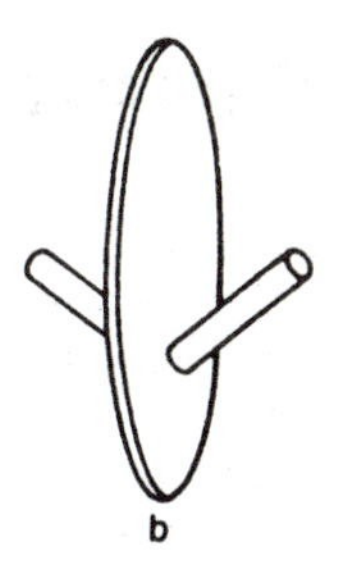

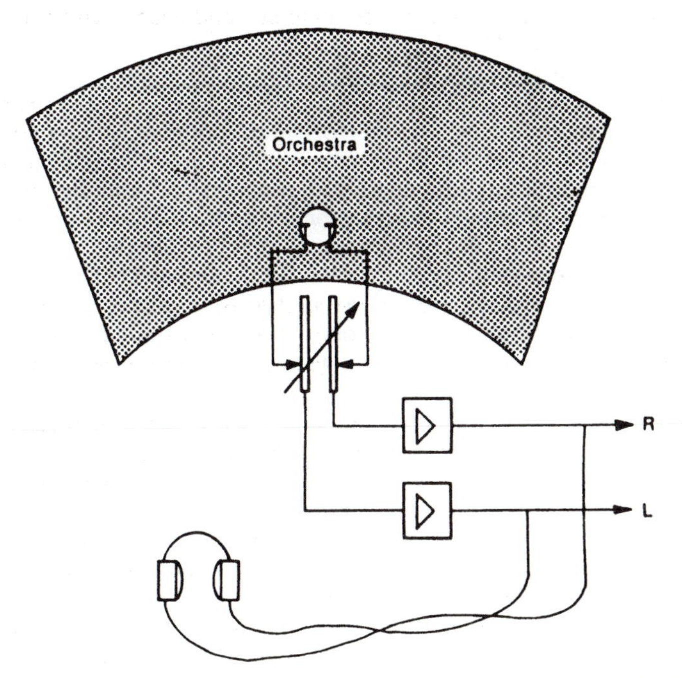

The microphone arrangement is placed in a position where the internal balance is optimum, e.g. above the conductor's head in an orchestra.

The two outputs must be kept separate, but in phase, throughout the recording/broadcast process so that they can be connected to the respective earpieces of a pair of headphones.

Stereophony is a method of obtaining spatial sound using loudspeakers.

Stereophony

On the previous page we discussed the realism that can be added to reproduction by listening binaurally, i.e. with two ears. There are two disadvantages in the system, however:

1. It requires the listener to wear headphones, which is not always convenient.
2. It requires a microphone technique that is only suited to sound sources that are well balanced within themselves, e.g. symphony orchestras and small classical ensembles, radio drama, etc.

Stereophony, on the other hand, is a two-channel sound system that is intended to be listened to via loudspeakers.

The basis is rather similar to the binaural system in that it sets out to provide the ears with separated sources and thereby supply information about the direction of the sound. Because loudspeakers are used, however, the separation is not complete. It is therefore common practice to make the positional information rather larger than life so that it is still effective even when diluted by the acoustics of the listening room.

Stereo listening conditions

To hear stereophony properly, the two loudspeakers should be perfectly matched, equally balanced for volume and in phase. Ideally they should be at least 2 m apart and an equal distance from the listener. There are various methods of checking the balance and phase of the loudspeakers. The simplest is to feed both with a monophonic source of sound (many stereo amplifiers are equipped with a stereo/mono switch) and to check that the sound appears to be located in the centre between the loudspeakers. If it appears to be to one side, or if it is not possible for the listener to be located at the same distance from each loudspeaker, it should be possible, within limits, to centralize the apparent position of the sound by adjusting the relative volume levels, either individually or with a 'balance' control. If the outputs are not in phase there will appear to be a hole in the middle and there will be a sensation that one's ears are pulling in different directions. A measure of the match of the equipment, especially the loudspeakers, can be judged by the apparent width of the mono source of sound, which should be as narrow and as clearly located as possible. Amplifier systems that provide separate control of frequency response (possibly graphic equalizers) on each channel require very careful adjustment to preserve effective stereophony. Special test records are available for these checks, but a good approximation can be obtained, using a radio tuner, by listening in the mono condition to the applause at the end of a concert. The sound of many hands clapping has a very broad frequency spectrum, very like the white noise (sound of equal intensity at all frequencies) used on some test records.

1. The optimum listening position for stereo is just outside the apex of an equilateral triangle with sides of about 2 to 2.5 metres. The diagram shows the difference in the path lengths for the two ears.

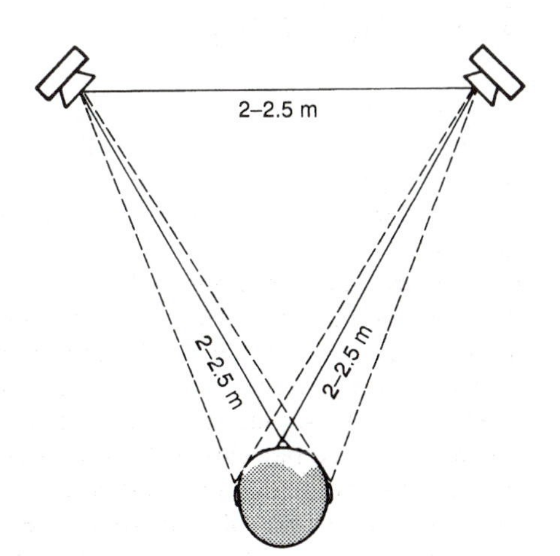

2. A domestic situation in which the loudspeakers are too widely spaced. The 'hole-in-the middle' effect can be mitigated by using a third loudspeaker, mounted centrally and fed with an attenuated addition of the left and right signals, adjusting the relative gain of the two amplifiers to centralize the sound image from the position of the listener.

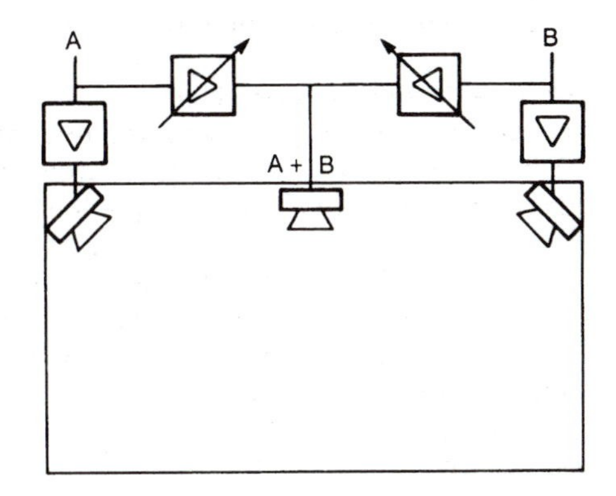

For effective stereophony the acoustics of the listening room must be taken into account. Obviously the room should not have too much reverberation as this will blur the ability to pinpoint the sources of the sound, although to a certain degree this factor is usually taken into account when the stereo records are made, i.e. the recording balance is monitored on loudspeakers in a room with some reverberation. Possibly even more important is that the acoustics of the room are reasonably symmetrical from the acoustic point of view. In many rooms this condition may be best obtained by placing the loudspeakers in corners of the room. This may result in the loudspeakers being too far apart, so that the stereo image has a hole in the middle. In this case the problem can usually be solved by the addition of a third loudspeaker, located between the two and fed with an attenuated mixture of the two stereo outputs.

Stereo sound produces an aural 'picture'.

Stereophonic Microphone Arrangements

There are two basic approaches to the production of stereophonic sound. There is the 'purist' approach, in which the intention is to create an aural 'picture' that agrees with the geographical locations of the sound sources, and the approach that takes no account of positional reality and arranges the aural images from a purely artistic viewpoint.

The 'purist' approach
The most straightforward technique involves using two directional microphones facing in different directions. These could be spaced apart (possibly to represent the spacing of our ears) or co-incident, as closely spaced as possible. Stereo microphones are available that incorporate two head capusules, set at different angles to each other. The capsule facing to the left provides the left-hand channel (destined for the left-hand loudspeaker) and the right the right-hand channel. In the case of the spaced microphones the discrimination between the channels results from a combination of the differing distances between the source and each microphone (propagation loss and phase difference) and the levels of output due to their directional characteristics. With co-incident microphones the discrimination is determined by the directional characteristics of the microphones alone. This is normally the preferred arrangement, both for convenience and the fact that as sound arrives simultaneously at each capsule, combining the two channels (e.g. to provide a mono output) does not result in time or phase distortion.

The A–B (or X–Y) axis system
If two figure-of-eight mics are set at ±45° to a sound source, the sound pick-up will be as in diagram 1 opposite. The left- and right-facing capsules provide the appropriate outputs; there is a dead area at each side and an equal live area at the rear where the L and R outputs are reversed. Working in this area must be avoided as it produces cancellations in mono and inaccurate positioning of the stereo image.

The M–S system
Another way to consider a stereo signal is in terms of an M (middle or mono) signal and an S (side or stereo) signal. The M signal represents the straight ahead or mono signal and the S the off-centre or positional information component. These signals can be derived from A and B outputs by adding A and B to produce M and subtracting B from A to provide the S signal. Microphones are available that produce M and S outputs directly. Others provide for both systems on a variable basis.

Some alternative configurations are illustrated opposite. The choice of system and microphone arrangements for various programme requirements is dealt with on pages 90 to 120.

The channels can be identified by a standard colour code: A = red, B = green, M = white and S = yellow.

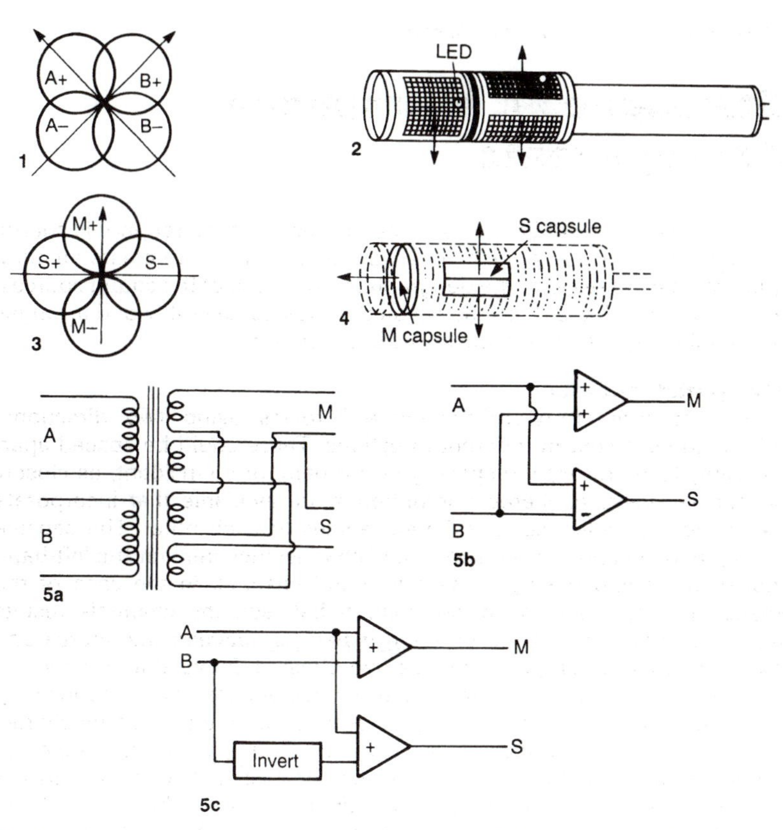

The basic principle of stereo microphones is possibly best illustrated by considering the directional characteristics of two figure-of-eight microphones arranged in two alternative configurations.

1. A–B (X–Y) axis. The front lobes of the microphone response (light gauze) provide the A = left and B = right stereo outputs. The rear (dark gauze) response is similar, but left to right reversed.

2. Stereo microphones for X–Y axis operation are usually arranged in side-working formation, with the ability to rotate one capsule with respect to the other. Some are provided with narrow-beam LEDs to help in identifying the angles from a distance when the microphone is slung high up.

3. M–S arrangement. This can be achieved with a side-working microphone by aligning the capsules at 90° to each other and rotating the whole microphone 45°, so that one capsule is pointing directly towards the sound source.

4. Microphones specifically designed for M–S working are usually 'end-fire'. They have the M capsule in line with their principal axis to provide the M (middle or mono) signal; the S capsule, for the S (side or stereo) complement, is set at right angles, across the width of the microphone.

5. A and B signals can be matrixed from M and S and vice versa. (a) Transformer matrix: the outputs are connected in phase (adding the outputs) for M and out of phase (subtracting the outputs) for S. (b) System using an amplifier with a summing input for the M signal and a subtracting input for the S signal. (c) Alternatively two summing amplifiers can be used, with one leg of the B (or S) signal phase-reversed.

Stereophonic Microphone Types

Stereophonic sound can be picked up using ordinary microphones, either separately ('pan-potted' into position in the stereo image; see p. 84) or arranged as coincident pairs, with their diaphragms as close together as possible. It is best to use a dedicated stereo microphone as it is important that the two microphones match each other exactly if a clear image is to be obtained.

The choice of stereophonic microphones depends upon the technique to be used and the sound field required, bearing in mind that in any application of microphones it is often just as important to consider sources of sound that are to be discriminated against as those that are wanted.

The diagrams opposite illustrate typical stereo microphone directional responses.

1. Crossed figure-of-eight microphones. The position giving the maximum output from microphone A (fully left) coincides with the zero response of the other and vice versa, so that the 'sound stage' subtends an angle of 90°. A similar response occurs at the back, but the left and right images are reversed.
2. Crossed cardioids provide a stereo image over an angle of 180°, with a diminishing response and no further increase in stereo width up to about 270°. As cardioids have no back lobes there is no out-of-phase area.
3. Crossed hypercardioid microphones provide an acceptance angle of about 130°, between cardioid and figure-of-eight, with a smaller out-of-phase area.
4. An alternative method of working is for the microphones to produce an M–S signal direct (this can then be matrixed to form A–B). This can be achieved with figure-of-eight microphones by rotating them through 45° and making the output from the A capsule provide the M signal and the other, at right angles, the S signal.
5. Cardioid and figure-of-eight configuration. With this type of microphone it is usual to have a cardioid capsule for the M signal, normally facing in line with the principal axis of the microphone (end fire), and a figure-of-eight capsule at right angles to provide the S signal. This arrangement provides a very useful method of adjusting the width of the stereo image by altering the relative levels of the two outputs. Varying the ratio of the S to M output alters the width of the sound image thereby creating a *zoom* effect. Maximum width occurs when the two outputs are equal.

A narrower minimum angle is achieved by incorporating an interference tube (as in a gun mic) in front of the M capsule.

The M–S configuration is particularly suited to television sound, where the ability to work on a boom at a distance from the sound source and vary the stereo angle is a requirement. Television sound coverage is discussed in detail in *Sound Techniques for Video and TV*, another Media Manual from Focal Press.

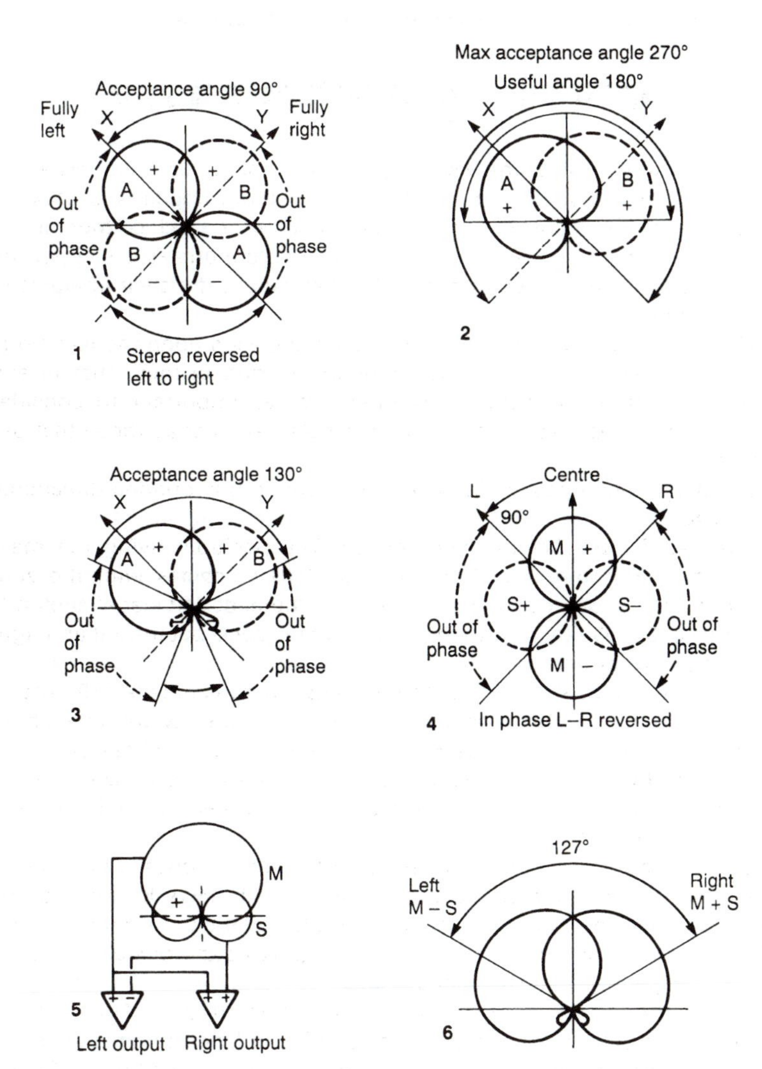

Typical stereo microphone directional responses (polar diagrams)

1. Crossed figure-of-eight microphones (usually side-working).

2. Crossed cardioid microphones provide a wider acceptance angle for the frontal sound field.

3. Crossed hypercardioid microphones have an acceptance angle between cardioid and figure-of-eight.

4. Two figure-of-eight microphones arranged for M–S working.

5. Cardioid and figure-of-eight combination. The cardioid element provides the M signal and the figure-of-eight the S signal.

6. Equivalent response characteristic of the above when the M and S outputs are added and subtracted in a matrix.

Stereophonic Sound Mixing

On page 80 it was stated that the 'purist' approach to stereophony is to erect a stereo pair of microphones in such a position that they produce a sound image of the source before them, which is relayed to two spaced loudspeakers. This can produce excellent results, with very clear positional information, but it provides little opportunity for the sound to be processed in the way that is normally required to satisfy the stringent balance, perspective and level control requirements of domestic listening. The majority of sound sources, particularly music, cannot be properly recorded without the assistance of additional microphones, carefully mixed in to promote artistic balance and aural perspective.

Spot-mic mixing
Where the orchestra, or other source of sound, is arranged in a suitable formation the best arrangement can be to use a coincident pair of microphones (or stereo microphone) to provide an overall picture of the source and then to add in subtle amounts of the outputs of some suitably placed 'spot-mics' to reinforce the level of those aspects that need 'spotlighting' to achieve artistic balance. To maintain the stereo image the outputs of these spot mics should be 'pan-potted' into a position in the sound stage that agrees with the aural picture presented by the overall stereo microphone. (The term pan-pot is explained in the glossary.) Ideally the spot microphones should be connected through variable delay equipment to make the timing of the sounds coincide at the stereo microphone. (Sound in air travels at about 0.34 m/ms (1 ft/ms).) Most modern sound control consoles incorporate at least one stereo microphone control, in addition to several line-level channels for connecting tape recorders, phonographs and external sources etc. These channels have ganged controls to enable volume control to be exercised without causing the sound image to shift. It is also important for ancillary controls, such as frequency response (EQ) and compressors etc., to be ganged if a sharp positional image is to be retained. If an M–S microphone system is in use provision should be made to adjust the level of the S signal relative to the M signal to alter the width of the stereo image.

Popular music balance
Most popular music balance involves multi-microphone technique (some rock groups use up to ten microphones on a drum kit!). There is no attempt to arrange the musicians in a 'panoramic' formation, in fact different sections may be screened off from each other. Much of the sound can come from electronic instruments which may be connected directly into the mixing console. Vocalists could be in a soundproof booth or recorded at a different time. The aural picture is therefore entirely a matter for artistic judgement and controlled by the pan-pots.

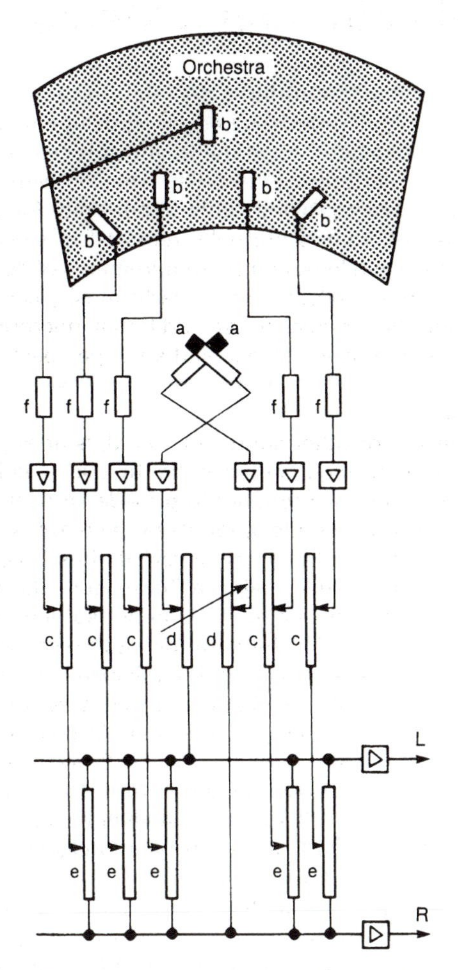

A schematic diagram of an orchestra, with a straightforward layout. The microphone balance consists of a coincident pair stereo microphone (a) and a number of 'spot' microphones (b). Each microphone is adjusted for level by the channel faders (c), the stereo pair being controlled by the ganged pair of faders (d); equalizers and channel amplifiers for the stereo channels should also be ganged.

The outputs of the spot mics are adjusted by the pan pots (e) to bring them into a geographic relationship that agrees with the positional information from the output of the stereo pair. Additional stereo microphones could be used in the balance, but care should always be taken to ensure that their apparent position agrees with the overall stereo picture.

Ideally, audio delays (f) should be inserted in the outputs of spot microphones to make the time of arrival of each source equal at the main stereo microphone.

Quadraphony provides sound with positional information for all four quadrants around the listener.

Quadraphony

The limitations of stereophonic sound

Stereophonic sound, discussed on the previous pages, can greatly enhance the realism of the reproduction and give the listener a measure of choice in balancing the various sources (i.e. instruments) and the studio ambience. It provides the listener with a small amount of two-dimensional discrimination but it is not the same as being there in person, when the acoustic ambience completely surrounds the listener.

A closer approach to reality can be achieved by using four channels with four loudspeakers arranged to supply a sound image for each of the four diagonal quadrants around the listener. In its simplest terms it can consist of two stereophonic systems arranged back to back.

Practical considerations

Quadraphony can be most effective when used to reproduce the normal 'concert hall' situation, i.e. with the orchestra in front and the ambience to the sides and rear. Unfortunately the improvement is rather subtle when it comes to persuading the public to buy a four-channel amplifier and four loudspeakers with their attendant problems of accommodation. The temptation is, therefore, to make the effect more startling by 'pan-potting' (p. 84) the sources so that they appear to be coming from all around the listener. Whether this effect is really desirable, with the possible exception of certain works specially written for spatial effect (e.g. Stockhausen's 'Gruppen'), is a matter of opinion but it is perhaps due to these sorts of considerations that quadraphony has not made much impact to date.

Another possible reason is the complexity of the equipment required to matrix four channels from the two-dimensional grooves of gramophone records. This is no problem for digital recording systems (e.g. CDs and R-DAT), which can provide four discrete channels.

Three-dimensional sound

Much of the advantage of quadraphony could be achieved with three channels. In fact it can be shown that all the horizontal information could be contained in three. The fourth channel could be used for the vertical dimension if there were no need to retain stereo compatability. As mentioned on page 18, although our hearing mechanism is designed to provide us with directional information mainly in the horizontal plane, we can also recognize the vertical element to a certain extent. This would make for realism but it would be a physically awkward format to accommodate in the average sitting room.

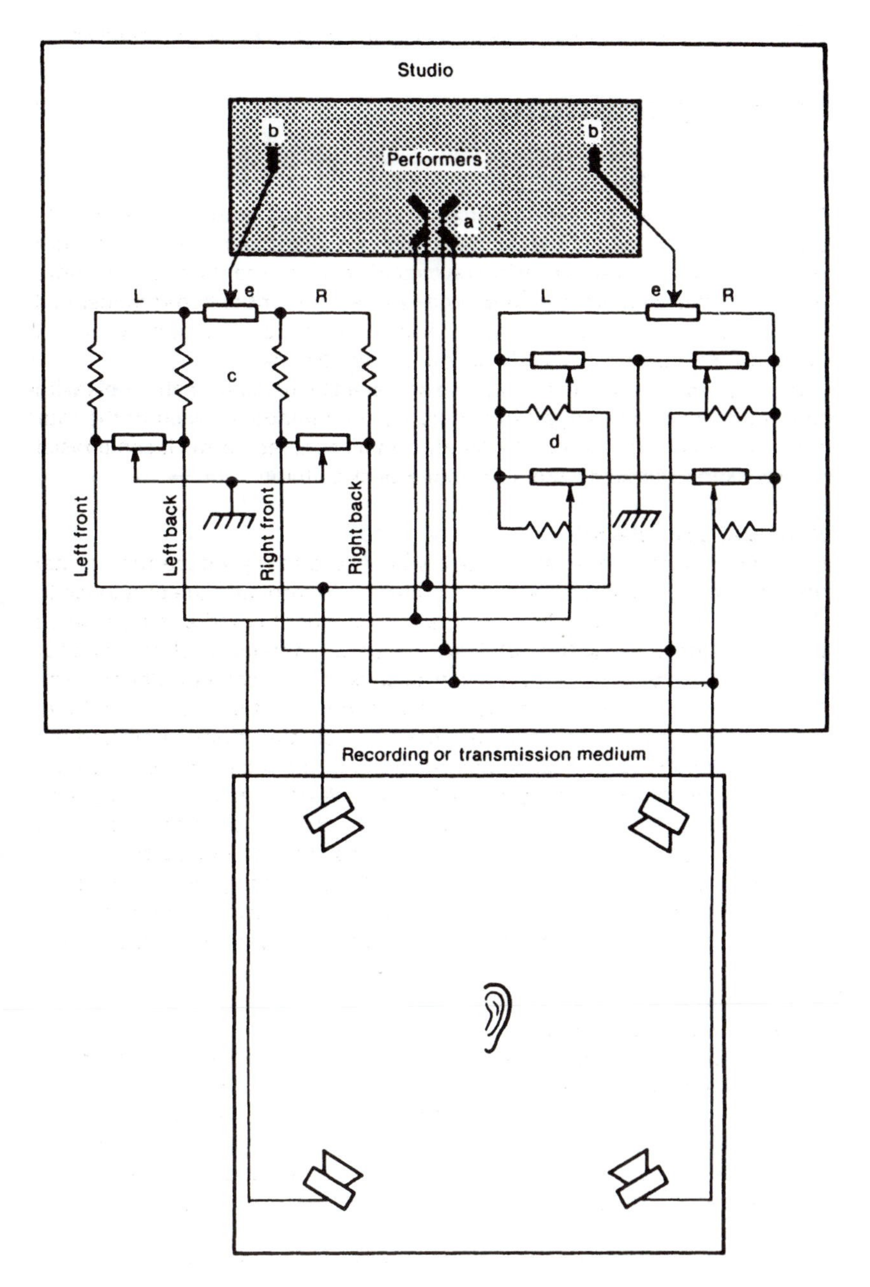

The performers can be grouped in front of or all around four microphones possibly arranged as two back-to-back stereo pairs (a). The outputs go eventually to four loudspeakers arranged on the four diagonal quadrants with respect to the listener. Additional 'spot' microphones (b) can be added to the balance and 'panned' in to the required apparent position. (c) and (d) are two types of front-to-back pan pots which can be fed from the stereo pan pots normally available in stereo consoles (e).

Using a system of 'matrixing' and four loudspeakers, it is possible to surround the listener with sound.

Surround Sound

The stereophonic effect

Conventional stereophony can provide a very effective aural image of a sound source but in only one plane. In the real-life situation, the listener in the concert hall would also be receiving reflected sounds from all around. In drama, dialogue and sound effects would come from all directions.

Ambisonic sound

The 'surround sound' effect can be simulated by matrixing (i.e. electronically combining in particular phase) two channels to provide outputs, containing the direct and ambient component, to supply additional loudspeakers. Decoders are available which generate signals for four or more loudspeakers in appropriate phase to suit their position. By combining three channels (or even two and a half channels because one need only have a bandwidth to 5 kHz) all three dimensions can be represented.

Sony Pro-Logic Surround Sound

In a two-loudspeaker system the central image is effectively a 'phantom' between L and R. The use of a central loudspeaker can make the listener's position less critical, but can result in a loss of separation between adjacent sources. To overcome this Sony Pro-Logic uses a complex active system of adaptive circuits. Separation is improved by deriving two internal control signals which continuously monitor the 'dominance' in either the L-R axis or M-S axis. When a dominance appears appropriate cancellation is applied to reduce the spill to adjacent loudspeakers, while maintaining the overall energy of the total signal. So when, for example, speech is dominant in the centre, that part of the signal is reduced by cancellation to the L and R speakers. (In a passive three-speaker system it would be present on all speakers.) When a wide signal appears (e.g. music) the L and R feeds are restored to the L and R speakers. With a complex signal (coming from all directions) the feeds are left intact. The S signal is restricted to 100 Hz–7 kHz.

Use of delay

The realism of the ambient (S) channel can be enhanced by introducing a delay of about 20 ms into the output, thereby exploiting the *Haas effect* (see p. 74).

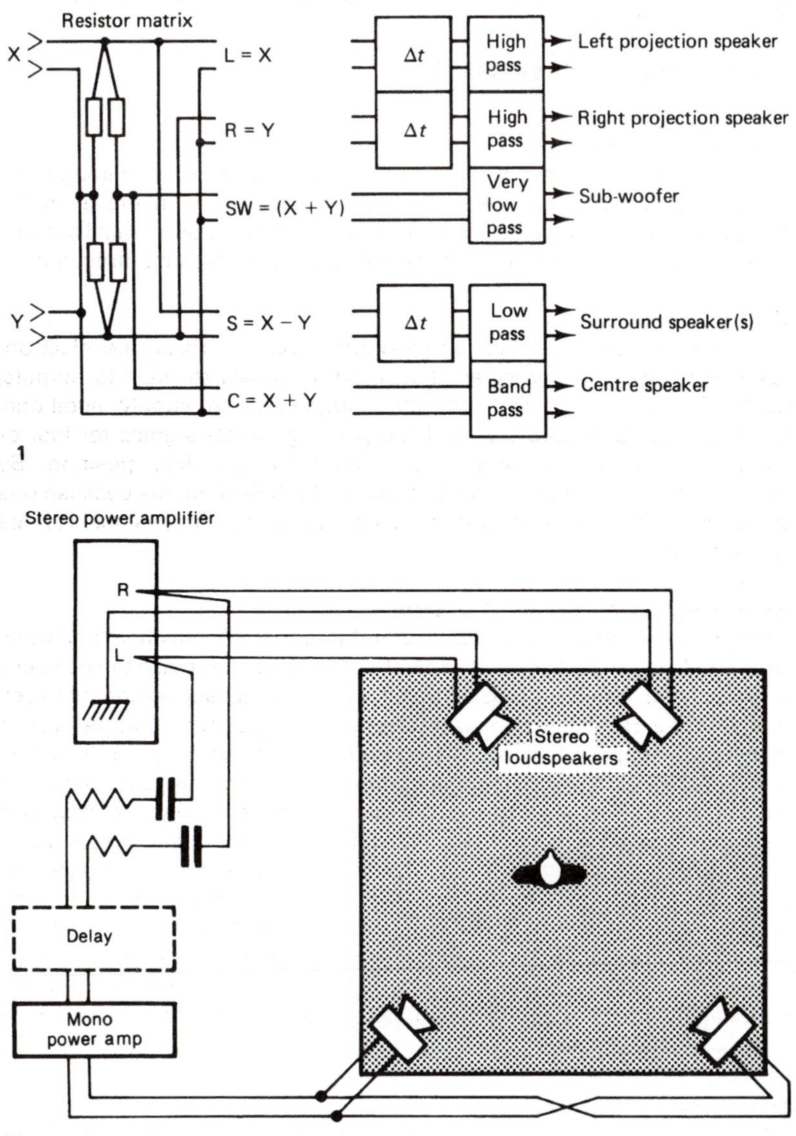

1. Basic Dolby surround sound decoding and typical processing arrangement. A new fast real-time processor can simulate sounds with variable distance and position.

2. A simple method of obtaining a form of surround sound. Most stereo power amplifiers have one leg of each loudspeaker connection common to earth. Connecting between the two live terminals therefore produces an antiphase output (R–L). This can be applied to a mono amplifier connected to two anti-phased loudspeakers.

Even simple speech recording requires careful consideration.

Balancing speech

When setting about making recordings or broadcasts of even a single speaker there are a number of factors to consider, all of which are interrelated. These include the acoustic situation, possible extraneous noise, the characteristics of the microphone and other equipment and the artistic effect, as well as the comfort of the performer.

Acoustic situation

Firstly it is important to consider the acoustic situation as it affects the comfort of the speaker. If the recording is to be made in a room it should ideally be of suitable size and have some reverberation (0.25–0.4 s is typical) so that the early reflections from the walls provide encouragement. Speaking in a very 'dead' room can be very tiring. On the other hand, to have too much reverberation can cause overlapping reflections to blur the diction and make speech less intelligible. This is particularly true if the room is large, so that even the early reflections are late. In this case it is advisable to surround the speaker as far as possible with some reflective surfaces (ideally acoustic screens).

Microphone positioning

Obviously the more reverberant the room, or the greater the risk of extraneous noise, the closer must be the microphone and the narrower its angle of acceptance. The best discrimination from ambient sound is obtained with a hypercardioid or shotgun microphone, although the latter type is not as effective against reverberation in a room as outdoors, because the random phase reflections tend to counteract the 'interference tube' principle.

Artistic effect

In general, the closer the microphone the more emphasis is put on diction, sibilants and nasal tones. The distance should be related to the loudness of the delivery. We tend to associate closeness and intimacy with a predominance of high frequencies. This can be achieved by working close to the microphone or, to a certain extent, by increasing the response (with equalization) in the so-called 'presence' region (around 3 kHz). On the other hand, working within a few inches of a directional microphone can produce a marked increase in bass response due to the spherical nature of the wavefront close to the mouth, increasing the difference in pressure between front and back of the diaphragm. This feature is often exploited by artists whose voices lack resonance. Working close to the microphone also increases the risk of explosive consonants (P-blasting). This is found to be worse at a distance of about 5 cm (2 inches). It can be mitigated by using a close-talking shield and speaking across the face of the microphone.

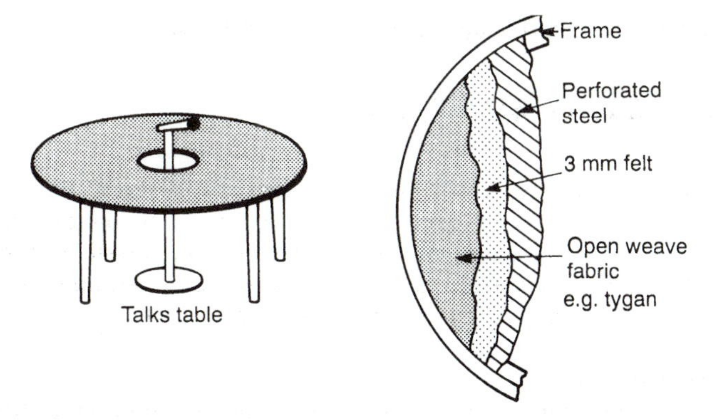

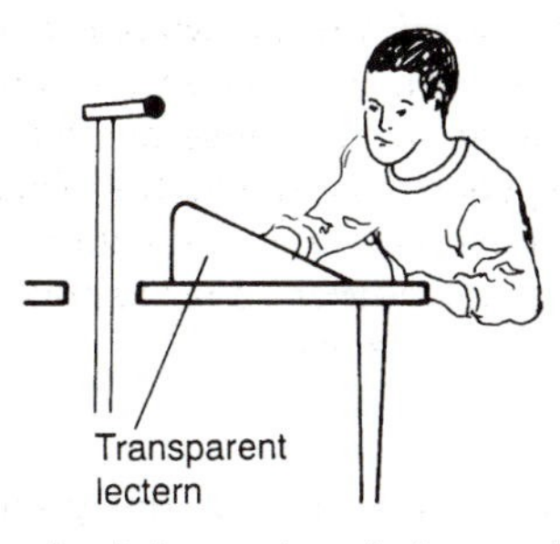

In the 'talks' situation it is usual to sit the speaker at a table. This should be designed to minimize reflections from the surface and reduce the possibility of movement noise. Special tables have been designed in which the surface is composed of fine mesh which is transparent to sound; these are usually provided with a hole in the centre through which microphones can be projected from floor stands and thereby not be affected by table thumping. Alternatively a solid-topped table should be covered by a thick cloth, preferably with a thin layer of plastic foam underneath.

An acoustically-transparent lectern can help to keep the speaker's head raised towards the microphone, but it can increase the possibility of script noises.

When two or more speakers are involved it might be tempting to use a single omnidirectional microphone in the middle of the group, but this is not to be recommended as it precludes the possibility of balancing the relative levels of the speakers. To arrange them at different distances from the microphone could result in different perspectives and their volume levels could vary considerably in the course of a discussion.

In the case of a broadcast in stereo it is important not to arrange speakers in a discussion at the fully left and right positions as, apart from rather overdoing the 'ping-pong' effect, it would present the mono listener with very little output.

There are numerous methods of covering the sound of people on the move. These are described in detail in *Sound Techniques for Video & TV*, another Media Manual from Focal Press. The simplest method is to use a hand-held mic, or a 'lavalier' (necklace) mic hung around the neck or clipped on to the speaker's clothes. This can result in unnatural and variable sound, rustling noises and changes of perspective as the speaker moves his head. It should be used in conjunction with a suitable equalizer, the output being compared with a regular microphone in a normal position and equalized to match.

Drama

Listeners to radio drama will be well aware of the ability of the sound medium to set the scene, conjure up images and create atmosphere without need of a picture. It is worth analysing how this is achieved, because it will always be an important factor, whether or not vision is involved.

Acoustics

Acoustic effect plays an important role in creating a sense of environment. Whether we realize it or not we associate the type and size of the room and the distance of the artist with the character of the reverberation, the shape of the decay curve and the timing of the initial reflections.

Earlier radio drama studios were constructed with a 'live' and a 'dead' end and had heavy double curtains that could be drawn across to separate them. These could be used to simulate indoor and outdoor acoustics. This is less necessary nowadays with the availability of digital artificial reverberation, variable delay and frequency shaping. A suitable studio is reasonably large (around 900 m^3, 30 000 ft^3) to accommodate a number of actors and allow them to move about, with a carpeted floor and a reverberation time of about 0.2 s. Outdoor scenes can be contained within a 'tented' area composed of sound-absorbing screens. The apparent size of a room can be suggested, using digital reverberation, by adjusting the timing of the first reflections and the shape of the decay. For example a small, reverberant room (e.g. bathroom) has large early reflections that build up the sound and decay fairly quickly, whereas in a large hall or church the initial reflections are delayed and of lower level but carry on for several seconds. The effect varies with the loudness of the initial sound. (If one whispers in church the acoustic is not noticeable, but shout and it is all too obvious.) The reverberation is the same. What changes is the threshold above which it becomes audible. Similarly, although distance is normally associated with the ratio between direct and reverberant sound, it is possible to distinguish between close-up sound in a reverberant situation and distant sound in a dry acoustic, even if the volume level is the same, by the shape of the decay and the frequency characteristic. For example, to simulate the sound of someone calling from a distance in the open air it will not work for the actor to call from the corner of the studio, as this exposes the studio acoustics. Rather he should work fairly near the microphone in the dead area, facing into the dead side of an acoustic screen, using little tone and with the bass and extreme high frequencies reduced by equalization. Before embarking on a drama production it is useful to experiment with various extreme situations in order to get the 'feel' of the studio.

Televised drama

The techniques involved in televised drama are very different from radio, because instead of the actors working to the microphones, in television

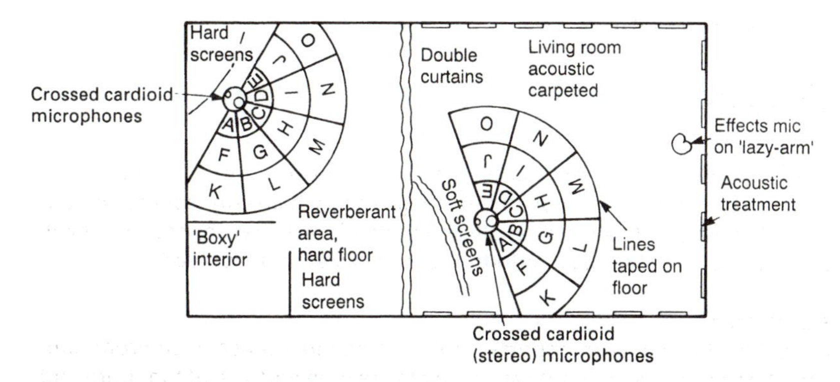

The choice and positioning of microphones for radio drama depend upon a number of factors, including the number of actors involved simultaneously, the required acoustics, convenience of operation etc. Microphones are usually mounted on stands. With a stereo setup the idea is to make the positioning of the artists relative to the microphone as natural and appropriate to the action as possible. For this purpose it is usual to mark lines on the studio floor which can be annotated in the script to remind the actors where they should stand for each speech.

the microphones have to follow the actions of the actors, while creating a sense of aural realism to reinforce the pictorial presentation. The sound techniques for televising drama are dealt with in detail in *Sound Techniques for Video & TV*, another Focal Press Media Manual.

Sound effects

Sound effects can be divided into two categories, recorded and 'spot'. Spot effects are sounds that are made in the studio, usually during the action of the play, by spot effects operators or by the actors themselves. The main advantage of spot effects over recorded sounds is that they help the actors to time their speeches, so it is usually best for the effects sounds to be made near the same microphone as the actors, unless the effect involves apparatus (such as a water sink) which is in a fixed position. The effects should be located or pan-potted into position to suit the action. Recorded effects should be 'folded-back' to the actors, either on loudspeakers or headphones, to enable them to pick up cues and sense atmospheres. Some sounds can be more effective when picked up indirectly from the foldback loudspeaker.

Atmospheric sounds should usually be spread over the whole stereo image. Portable digital recorders make it possible to collect authentic sounds, and digital hard-disc devices make it possible to store large numbers of sounds which can be cued instantly and in random order. See page 222.

Microphone positions for balancing a piano depend on the quality of the instrument and the type of music being played.

Recording the Concert Piano

Tone production

The piano is a percussion instrument. When a key is depressed a felt-covered hammer hits a string and immediately springs away, allowing the string to vibrate. The string vibrates at its fundamental frequency, determined by its weight, length and tension, and also at a number of harmonic frequencies which are multiples of the fundamental note. Some of these are discordant with the fundamental and have to be suppressed. This is achieved by giving the hammers a fairly broad surface and covering them with soft felt and arranging for them to hit the string at a point that damps the most discordant harmonics, notably the seventh and the very high 'clang' tones. If the felt has hardened or the keys are struck very hard so that the felt becomes compressed the upper harmonics are excited and the piano has a harsh tone.

Energy distribution

The treble strings are so thin that they make little sound on their own. They are made to pass over a bridge which transmits the vibrations to a sound board from which most of the tone is radiated.

The sound board is less effective for the very high harmonics which promote brilliance and attack. The nearer the microphone is to the strings the more brilliant is the tone. The bass strings, being thicker, radiate quite strongly at right angles to their length. One way of discriminating against a too heavy bass in a 'distant' balance is to place the microphone towards the tail end of the piano, looking towards the keyboard.

Choice of microphone position

The microphone position for a piano is determined by:

1. The ratio of attack to warmth of tone required to suit the music and character of the performance.
2. The tone of the instrument and its internal balance.
3. The acoustic conditions and possibly the need for separation from other instruments or noise.

The lid should always be open or off, otherwise the piano will produce a rather 'stifled' sound. In general the best results are obtained by positioning the microphone so that it can 'see' all the strings at a distance determined by the ratio of attack to warmth of tone required. This can usually be achieved along a line running at about 45° in each plane from the strings.

Stereo effect

If the piano is solo it might be desirable to provide a certain amount of treble/bass divide across the sound stage, using crossed cardioid microphones. In general, and certainly when the piano is in concert with other instruments, it is better to use an M–S approach, to provide a narrow image with a spread ambience.

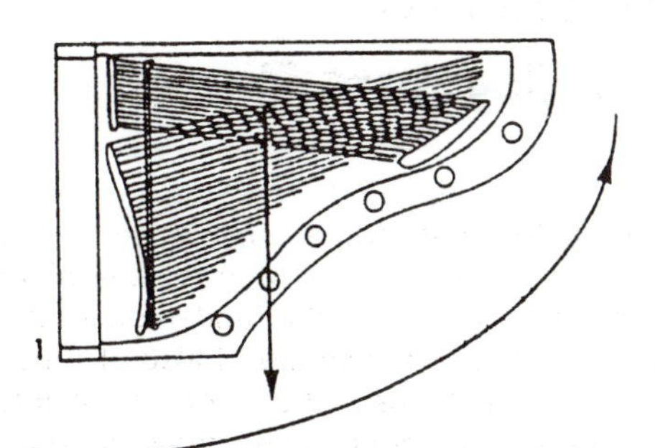

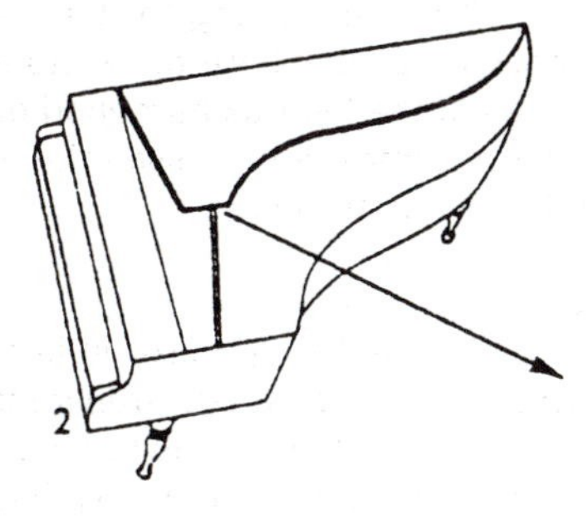

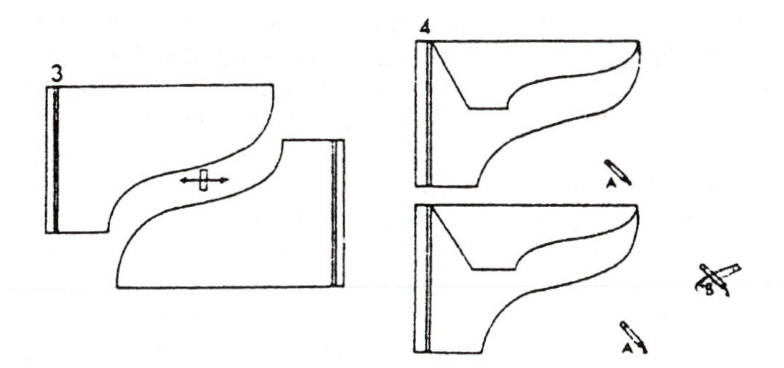

Distant balance for a concert grand piano.

1. A concert grand piano can have considerable bass radiation at right-angles to the strings. A more even balance can often be obtained with a microphone at about four metres distance by moving it towards the tail of the piano.

2. Best results are usually obtained with a microphone on a line diagonal to its three dimensions, so that it can 'see' all the strings.

3. Two pianos 'wedded' for a duet with the lids removed. Suitable microphone arrangements would be an M–S stereo mic suspended above, or two opposing cardioids either as a stereo pair or slung separately over each piano.

4. Two pianos in line formation. The balance could consist of two microphones as at 'A' pan-potted into position, or a stereo pair in position 'B'.

The rhythm piano requires a very close microphone at the treble end of the instrument.

Recording the Rhythm Piano

The technique for broadcasting the piano in light music and dance bands is quite different from that used for classical music.

The piano in a rhythm band

Traditionally the piano was part of the rhythm section of the dance band, its function being mainly to provide rhythm chords, filling-in phrases (mostly in the treble) and occasionally the melody. Modern orchestration tends to be rather more complex but the requirement remains for a very crisp, brittle tone (particularly at the top end) and good separation from the rest of the band so that, even when played quietly, it can be heard in competition with the much louder brass and saxophone sections.

Microphone techniques

The necessary tone is best achieved by using a unidirectional (cardioid) microphone pointing either almost vertically towards the top strings at a distance of about 20 cm (8 in) or pointing straight down into one of the holes in the iron frame (normally the second from the treble end) and only slightly above the frame so that it looks straight at the soundboard. This latter position does not produce quite such a crisp tone as the microphone pointing at the treble strings, but it can produce a remarkably well-balanced result. The reason for this is probably due to the fact that, although the sound board is designed to radiate all frequencies in phase, there is inevitably some antiphase cancellation at the upper end of the scale where the wavelengths are small. Concentrating on a small area tends to eliminate this effect.

A very neat arrangement can be made for either application by using an electrostatic microphone with short extension tube as a table stand resting on a foam rubber pad inside the piano.

The best mic position for most upright pianos is with the top lid or front cover off pointing at the top strings.

Multitone and electronic pianos

Some upright pianos have a third pedal (in addition to the sustaining and soft pedals) which causes a set of little cloth strips fitted with metal ends to interpose between the hammers and the strings. This greatly increases the harmonic content of the sound and produces a metallic tone, slightly like a harpsichord. Multitone pianos have been superseded by electronic keyboards with which a wide variety of tone colours can be obtained. These are usually picked up by a microphone placed in front of their cabinet (loudspeaker), or by direct injection into the sound mixer from a DI box.

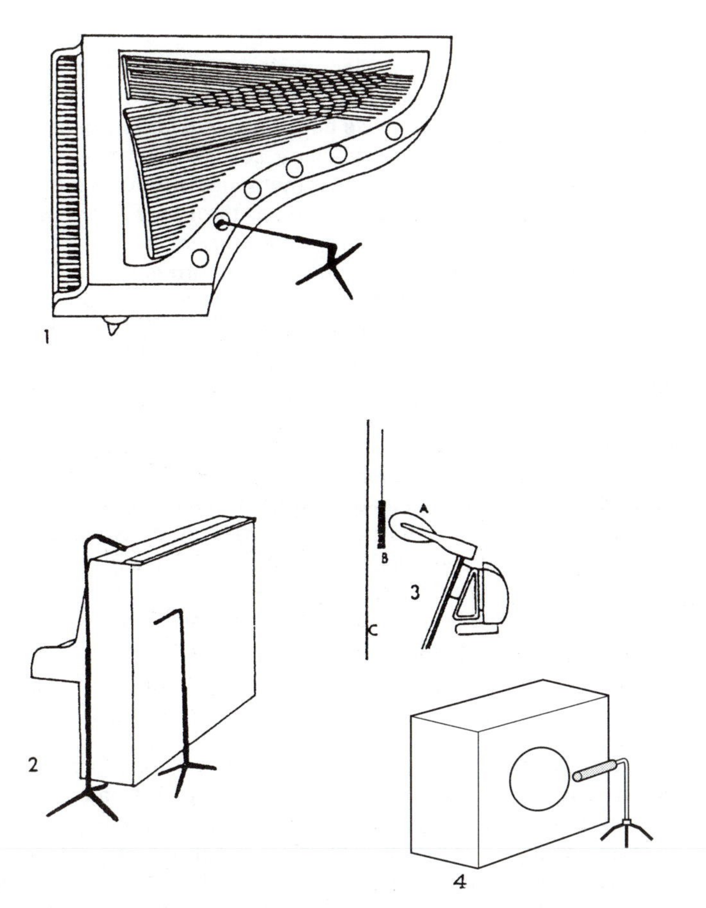

1. The microphone head would be just above the lip of the hole in the frame. A neater arrangement can be made by resting a short stand on the piano frame or clipping it inside the case.

2. The best arrangement is with the microphone over the open lid at the treble end. Alternatively, it can be placed behind the sound board.

3. *Arrangement for multitone piano.* A, felted hammer. B, cloth tab with metal end. C, piano strings.

4. With an electronic keyboard it is merely necessary to position a microphone (usually cardioid) in front of the loudspeaker, or take a direct output if available.

Instruments with complex harmonics sound louder than their power would suggest.

Various Keyboard Instruments

Harpsichord

The harpsichord is rather like the piano except that the strings are plucked with quills instead of struck with hammers. This produces a small tone that is very rich in harmonics. These harmonics are largely centred on the frequencies where the ear is particularly sensitive, so the harpsichord sounds much louder than a programme meter measuring its output would lead one to believe. A particular feature of the harpsichord is its loud action noise, which stems directly from the plucking action.

The best method of broadcasting the harpsichord is usually to employ two microphones, one underneath pointing towards the soundboard, but angled to discriminate against any noise from the pedals, and another microphone above in a similar but rather nearer position than that normally used for a classical grand piano balance.

Clavichord

The clavichord is a smaller, simpler precursor of the piano. Normally in the shape of a rectangular box, it sits on a table. The strings are struck with a brass tangent which remains in contact with them while the key is depressed so that vibrato (small rhythmic changes of pitch) can be applied by varying the pressure on the keys with the fingers. The tone produced is very small and delicate and calls for a very close balance unless the instrument is completely solo; this is best achieved with a microphone overhead and to the right.

Celeste

The celeste is a keyboard instrument in which the tone is produced by bells. Most of the sound comes out of the back, which has an open fret covered with thin fabric. Celestes vary considerably in purity of tone and power. They usually sound best with the microphone about two metres (six feet) away from and slightly above the back, but the need for separation from other instruments usually dictates a much closer balance, often about 20 cm (8 in) from the back.

Organ

Most of the organs encountered nowadays are electric. Their sound comes out of loudspeakers either contained within the console or in separate units. It is possible in many cases to take a direct feed from the organ to the sound mixing desk but, unless the organ is properly adapted for this purpose, there may be matching problems and it is simpler to place a microphone in front of the loudspeaker.

Pipe organs can present the problem that they 'speak' from a number of different directions or even locations. Usually the best approach is to use omnidirectional mics close to each section to combine clarity with ambience. They can be pan-potted to provide a spread effect.

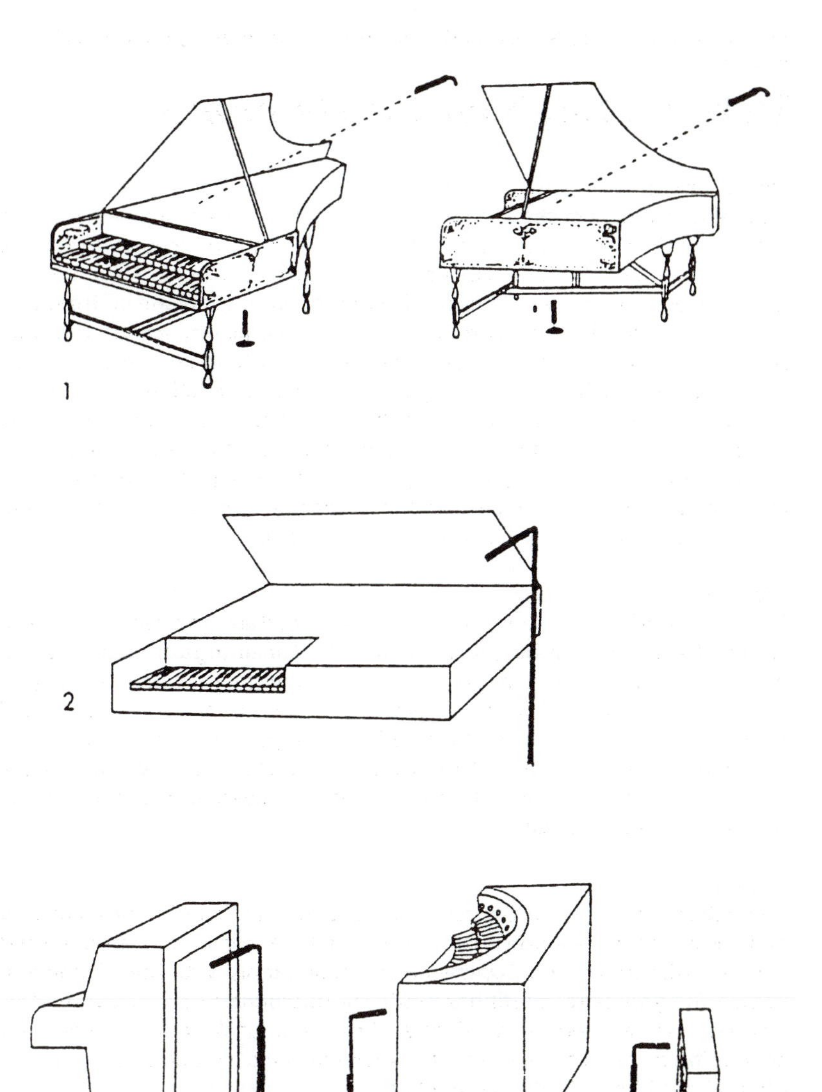

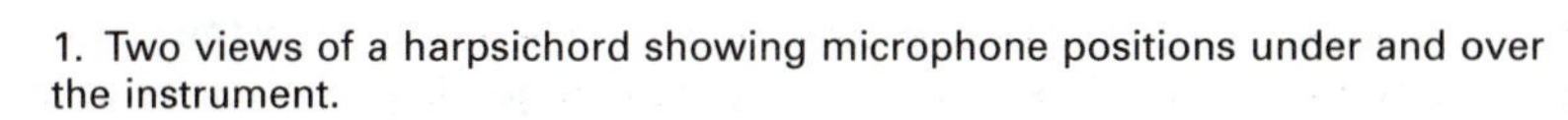

1. Two views of a harpsichord showing microphone positions under and over the instrument.

2. Clavichord, showing suitable microphone position.

3. Celeste, with microphone at the back.

4. Electronic organ with alternative position for microphone for internal loudspeaker and external loudspeaker.

Stringed instruments sound best with the microphone roughly perpendicular to their front faces.

Strings

The 'family' of strings used in most orchestras consists of violins, violas, cellos and double basses.

Violin

Most of the tone of the violin is radiated from the front face and from the 'f' holes. The back, which is connected to the front by a peg called the sound post, also makes a powerful contribution to the tone.

The high harmonics which give brilliance to the tone radiate in a narrow lobe at right angles to the face of the instrument.

If a fairly distant microphone technique is used, say 2 metres (6 feet) or over, the best sound is produced along this line. If, however, a very close technique is necessary to give separation from other instruments it may be better to work somewhat off axis, otherwise the tone can be rather 'whiskery' and variable if the violinist sways about.

The violins in an orchestra sit in pairs (called desks) facing the conductor, sharing a music stand. They are usually divided into two sections of 'firsts' and 'seconds' or, in some types of light music, three groups, A, B and C.

Viola

The viola is a larger version of the violin. It is tuned a musical fifth lower. It has thicker strings and produces a less brilliant tone which is not quite as directional as the violin.

Cello

The cello is tuned an octave lower than the violin. The strings are not unduly thick in relation to its size and it is fairly directional in its upper register. Cellos, or more properly, in the plural, celli, are also arranged in pairs in the orchestra, sharing a music stand between two.

Double bass

As played in the 'straight' orchestra the double bass does not constitute much of a problem as it is not very directional, but it is sometimes necessary to use a fairly close microphone technique to obtain clarity.

Rostrum resonance

Both the cello and the double bass are connected to the floor by the spike on which they rest. This means that the floor acts as an extension of the instrument, and if it is a rostrum great care is necessary to ensure that it does not resonate. This is achieved by making a false top with hardboard laid on felt and providing stiff bracing inside. In general the basses sound better on the studio floor.

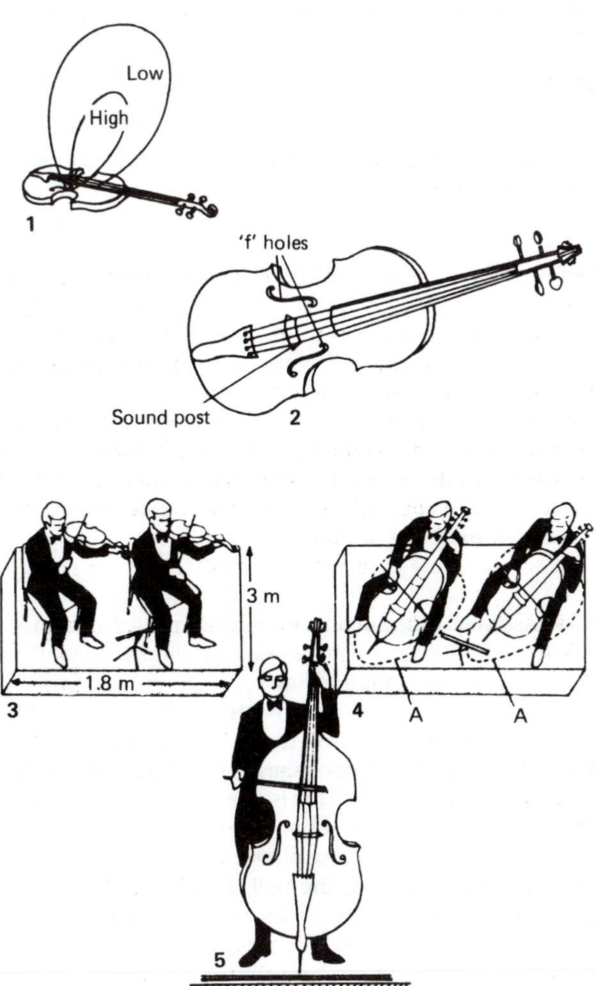

1. The violin radiates its high tones and harmonics in a fairly narrow angle slightly forward of normal to the front face. The low frequencies occupy a much broader field.

2. General appearance of violin showing position of sound post and 'f' holes. The sweetest tone (from within the resonating chamber) comes out of these holes.

3. One 'desk' of violins or violas occupies a rostrum space of approximately 1.8 × 1.3 metres (6 × 4 ft).

4. Desk of celli needs at least 2 × 1.8 metres (6.5 × 6 ft) unless the musicians sit diagonally. If possible 'cello boards', A, should be used.

5. When basses are used on rostra care must be taken to avoid resonance. The rostrum must be well braced and, if possible, a 'sandwich' of felt and extra top section should be used. A pair of double basses occupies a space of about 2 × 1.5 metres (6.5 × 5 ft).

Woodwind Instruments

In deciding where to place microphones for the woodwind there are two different approaches according to whether the music is 'straight' (i.e. orchestrated for a natural balance in the concert hall) or designed for microphone assistance with little if any attempt at internal balance. Most modern and all 'pop' music comes into the latter category.

Flute and piccolo

The flute and piccolo produce a very pure tone when heard from a distance, i.e. the microphone at least 2 metres (6 feet) away. At close range they have a 'breathy' quality which is often exploited in popular music to produce a crisp effect. In the ultimate the microphone is placed a few centimetres from the mouthpiece so that the breath puffs give each note a sharp leading edge. A special microphone is available which fits into the end of the flute for extra separation.

Oboe

The oboe has a sweet but piercing tone due to the fact that it is rich in harmonics, most of which are centred on frequencies to which the ear is particularly sensitive. The sound that emerges in line with the bell is particularly strident but this is directed towards the floor. Microphone distances of less than one metre (three feet) are seldom used. The instrument sounds best from a considerable distance. The same is true of the cor anglais (or English horn). This is a larger instrument with a natural key of F, a fourth lower than the oboe. It has a warm, complex tone often used to suggest a 'pastoral' atmosphere in programme music. This is created by the unusual shape of the bell, which is slightly spherical and closed in somewhat at the end.

Clarinet

The clarinet has a variety of different qualities according to the register in which it is played and the way it is blown. In the high register it can be very piercing. It can also be played 'sub tone' (below its fundamental register), when it sounds very soft and requires a very close microphone technique.

Because of its many registers (usually divided into 'medium', 'chalumeau', 'acute' and 'high') it is difficult to play across the breaks between. For this reason clarinets are made in several pitches, the ones in most common use being B flat and A. Military bands often contain an E flat clarinet, which has such a piercing tone it is capable of cutting through the rest of the band.

Bassoon

The bassoons are normally only used in straight orchestras. Their sound is not very directional so they pose few balance problems.

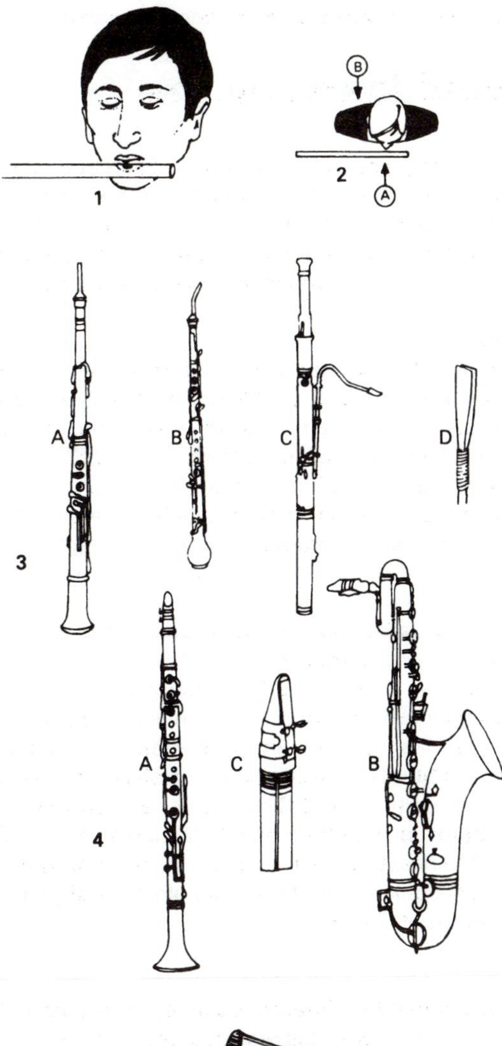

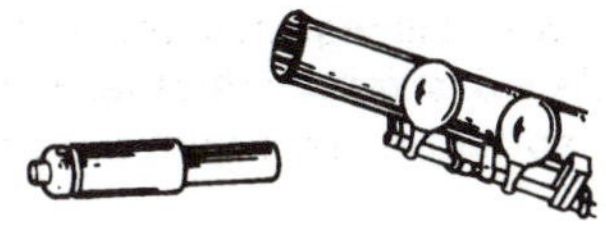

1. The tone of the flute is produced by blowing over a small aperture. (It is said that 'you don't play the flute, you stand beside it and pop the tune into it'.)

2. A crisp tone is produced with a microphone in front of the mouthpiece (position A; a purer tone from position B).

3. A, oboe, B, cor anglais and C, bassoon use double reeds. D, double reeds are two reeds fixed parallel on either side of the mouth tube, leaving a narrow gap between.

4. A, clarinets and B, saxophones use single reeds. C, single reeds form a narrow gap with the mouthpiece.

5. Flute pick-up cartridge.

Mutes change the sound quality, volume and directivity of brass instruments.

Brass Instruments

The brass instruments require a different approach from the woodwind by virtue of their greater power and directional properties.

Trumpets and trombones

Trumpets and trombones project their sound most powerfully in line with their bells. The high harmonic overtones that give brilliance are confined to quite a narrow angle so that if the microphone is right on axis the sound quality varies considerably if the player changes direction. It is, therefore, invariably better to work slightly to one side of or over the direct line, depending upon visual considerations.

Visual considerations

When planning the layout for trombones it is necessary to allow space for the extension of the slides. It is also worth considering that the trombones make a spectacular effect when viewed from the side with their slides moving in unison, so when television is involved they should be arranged in a straight line that is accessible to the cameras.

Mutes

Various mutes are available for the brass. These alter the tone and reduce the volume in various degrees. In general the straight conical mutes produce a hard tone that is not very much less in volume than the open bell (if a microphone is not in direct line). The cup mutes that cover the bell reduce the volume very considerably and produce a softer tone. All mutes, especially the cups, tend to make the instrument much less directional so that it is satisfactory to work across the microphone.

Horns

The horns require special care in positioning because they point backward almost directly behind the performer so that the surface immediately behind plays an important part in the production of tone. It should provide reflection, but no resonance, and must not be so close that the performer is inhibited from playing at the proper volume.

Euphoniums and tubas

Euphoniums and tubas are mainly used in brass bands and are not very directional; they tend to sound best from an overhead position.

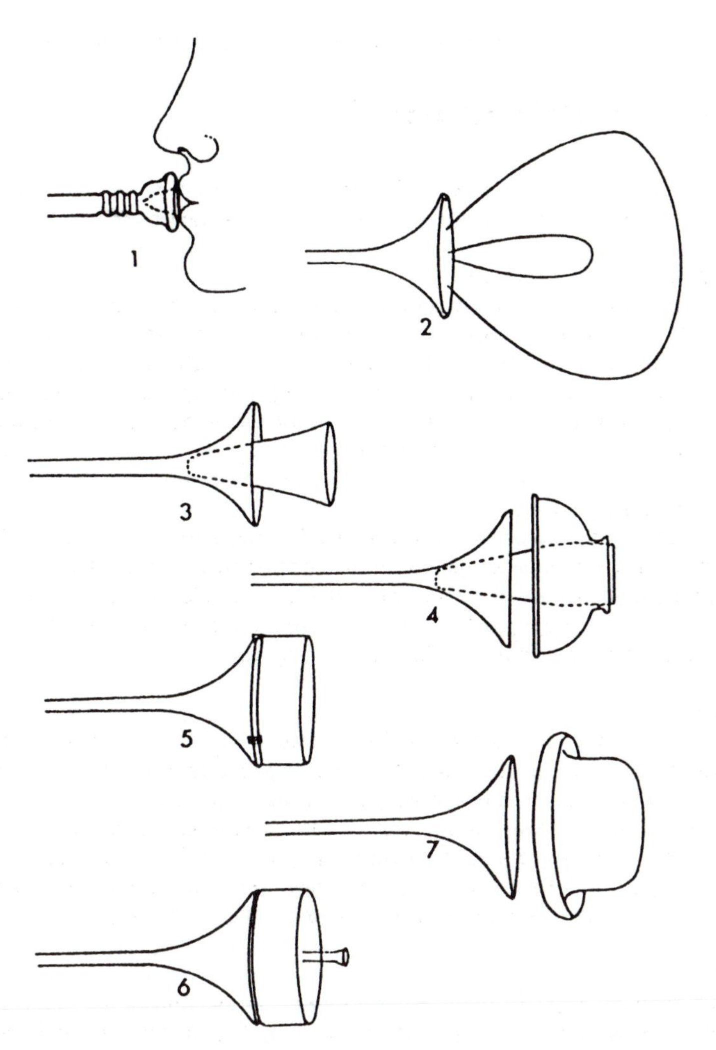

1. In the brass instruments the tone is produced by pressing the lips against a cup-shaped mouthpiece so that they form a pair of reeds, the air flow being controlled by the tongue.

2. Because of the shape, most brass instruments, particularly the trumpet, produce their most brilliant tone in direct line with their bells (the narrow lobe represents the HF response). Various mutes are used with trumpets and trombines (and to a lesser extent horns).

3. The straight mute, which produces a hard tone suggesting distance.

4. The cup mute – a soft, smooth tone.

5. The hush mute – a can with absorbent material inside clips over the bell to produce a very soft tone.

6. The Harman or 'wow wow' mute for shrill 'talking' effect (made by covering the end with the hand).

7. Bowler hat for swell effect.

Chamber Orchestra

The term chamber orchestra suggests a small combination of classical musicians playing music that was written to be played in fairly intimate surroundings. It may be a quartet or a small orchestra of the type for which much of Mozart's work was written. In any case the chances are that there will be a conventional layout which gives good internal balance.

String quartet

The string quartet has a conventional layout designed to help the musicians to play in concert and balance with each other.

In this type of music the parts are woven together in such a way that sometimes all make an equal contribution to the texture and sometimes each comes into prominence in turn. It should be possible to find a position for the microphone, usually about 2 to 3 metres (6 to 10 feet) in front and above, where these conditions are met. If a piano or harpsichord is added another mic will be required. If the sound is in stereo a stereo pair of microphones is required in the main position.

Classical orchestra

The type of orchestra used by most of the great composers up to about 1800 consisted of about 24 strings in the form 8-6-4-4-2 (first violins, second violins, violas, celli and basses), double woodwind, i.e. two of each (flutes, oboes, clarinets and bassoons), four or five brass (two trumpets and three trombones), two or four horns, and timpani.

The standard layout, comprising a semicircle, with the strings in front, woodwind and finally brass and percussion behind, makes a logical combination that should balance on one microphone (two if stereo) somewhere above and behind the conductor's head. However, it is usually necessary to add at least one other microphone to accentuate the woodwind and give extra clarity and better perspective.

The procedure is to experiment to find the best position for balance and at the same time the best compromise between clarity (nearness) and warmth of tone (reverberation), then to listen critically to the rehearsal and tell the conductor of any points where the balance is not satisfactory. He will expect this sort of guidance and can usually make the necessary correction.

In very reverberant conditions it may be necessary to close in and employ more microphones, each 'pan-potted' into position in the stereo image. If the presentation involves video or televison it usually requires a closer approach to give the sound a clarity of attack that will complement close-up pictures of the musicians. This aspect is discussed fully in *Sound Techniques for Video & TV*, another Focal Press Media Manual.

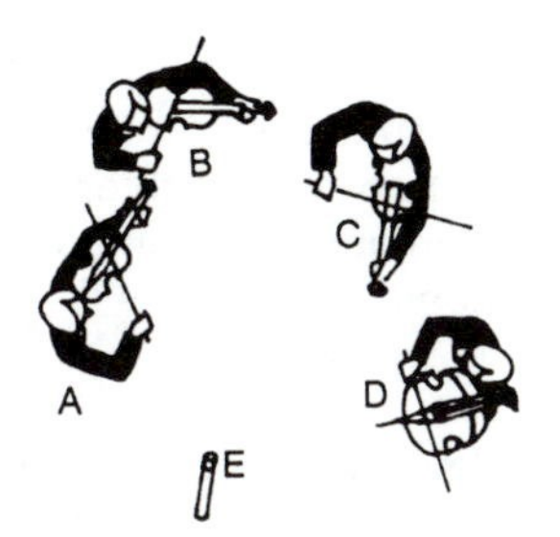

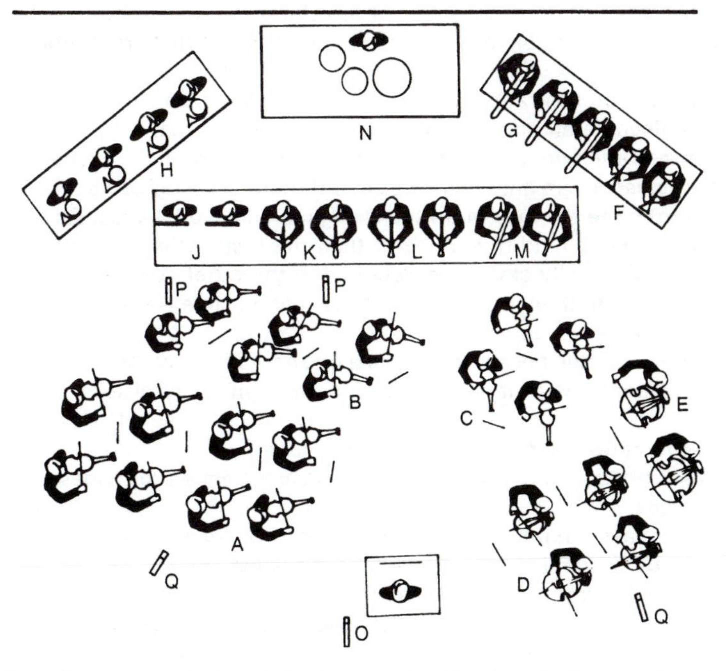

1. The string quartet. A, first violin. B, second violin. C, viola. D, cello. E, typical position for microphone at about 2–3 metres (6–10 ft) high.

2. The classical orchestra. A, first violins. B, second violins. C, violas. D, celli. E, double basses. F, trumpets. G, trombones. H, horns. J, flutes. K, oboes. L, clarinets. M, bassoons. N, timpani and percussion. O, possible position for a single microphone balance. P, position for woodwind accentuation microphones about 2–3 metres (6–10 ft) high. For stereo the main microphone would be a stereo pair, augmented by at least two others Q, 'pan-potted' into position in the stereo image.

Light orchestras require multi-microphone technique for balance and acoustic effect.

The Light Orchestra

Light orchestras come in all sizes and combinations and play all types of music from light classical to pop. It is, therefore, not possible to generalize as regards their balance technique except in one respect: they are all likely to require a multi-microphone balance.

Light music balance

A typical combination to play straightforward light music would consist of about 22 strings (typically 8 first violins, 6 second violins, 4 violas, 2 celli and 1 or 2 basses), 3 trumpets, 2 or 3 trombones, 2 horns, double woodwind (i.e. 2 of each: flute, oboe, clarinet and bassoon), some of whom may also double on saxophone, piano, possibly harp, timpani and percussion.

Even if the orchestrations are written for internal balance this is unlikely to be achievable on one microphone. It will be seen that the power of the wind, brass and percussion is very similar to the symphony orchestra but the string section is only half its size. It therefore becomes necessary for the microphone to close in on the strings to achieve balance. As a result the strings will sound much nearer than the rest of the orchestra so we have to add equally close microphones to the other sections, fading up just enough of their output to restore perspective while maintaining balance.

Once it becomes necessary to use a close microphone no section must be left without one or it will be missing, or sound very distant.

Multi-microphone techniques

The use of separate microphones for each section provides several advantages:

1. Time is saved in finding the optimum position for a single microphone. This can be a premium in the rehearsal situation.
2. The music arranger is freed from the requirement to consider internal balance and can write for all sorts of novel effects that exploit microphone techniques.
3. Greater freedom of layout. The orchestra can be arranged in a spread formation to help separation or for the sake of appearances.
4. The use of close microphones with added artificial reverberation produces a 'crisp', brilliant tone which suits modern orchestration and is congruous with close-up shots of the musicians.
5. Various forms of enhancement, e.g. tonal modification, can be applied individually.

When an overall stereo microphone is in use the accentuation microphones must be pan-potted into positions to match the stereo image.

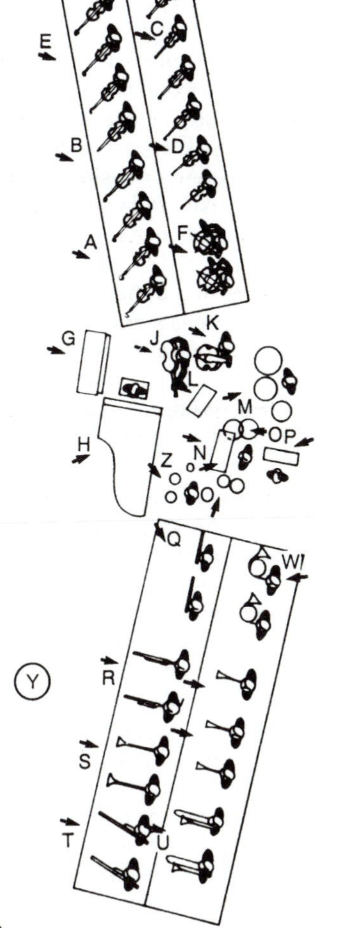

1. Typical 'straight' layout for a light orchestra, with possible positions for accentuation microphones as outlined below.

2. Spread-out split layout for visual effect. Because of the unnatural spread of the instruments, and possibly also to balance for modern orchestrations, a large number of microphones, each at fairly close range, is required.

The microphones (pictured here as arrows) accommodate the following sections: Microphones A, B, C, D, violins. E, violas, F, celli. G, celeste. H, piano. J, acoustic guitar. K, double bass. L, electric guitar amplifier. M, timpani. N, bass drum. O, side percussion. P, vibraphone. Q, flute. R, oboes. S, clarinets. T, bassoons. U, trombones. V, trumpets. W, horns (they project their sound backward). X, overall string microphone. Y, overall woodwind and brass microphone. Z, bongoes. AA, accordion. BB, bass guitar.

X and Y can be very useful to restore cohesion which may be lost with very close microphone technique. They also tend to bridge the gap between the direct sound and the added reverberation when an acoustically dead television studio is used.

Symphony orchestras invite a rather 'purist' balance technique.

The Symphony Orchestra

The symphony orchestra can be composed of about 100 musicians, according to the music being played. A typical combination would consist of about 66 strings, in the form of 20-16-12-10-8 (using the notation explained on page 106), triple woodwind, seven brass, including bass trombone and tuba, four horns and at least two percussion players.

Internal balance

Most symphonic music is written to produce an internal balance and an even perspective from a position well back in the hall, and the composition of the various sections and their relative positions is designed to this end. A fairly distant perspective is usually required, with a warmth of tone engendered by the hall acoustics, commensurate with the character of the music being played. The ideal, therefore, would be to record the whole orchestra on one microphone, or stereo pair, somewhere in the vicinity of the conductor's head and leave the balancing to him. In practice this would seldom work. First, the conductor is not ideally sited to judge the balance with sufficient precision to satisfy the stringent requirements of recording (with its restricted volume range and normally reduced level) so, whatever system of balance is employed, the conductor expects to receive guidance from the sound control room. Second, this may not be the ideal position for a simple microphone balance. Nevertheless the use of a single stereo microphone is a good starting point from which to add 'spot' microphones to the balance as necessary.

Overall balance

In deciding where to place the main overall microphones a number of factors have to be taken into account, some of which work against each other. They include: sectional balance, perspective, ratio of direct to reverberant sound, clarity and attack. The diagram opposite shows the various factors that influence the positioning of the main microphones, the direction of movement required for each effect and the area in which all the factors might come together, based on the author's experience of balancing many large orchestras in a variety of halls and studios. In general it is an indication of a good acoustic that there is a large area in which these factors overlap. If the acoustics are not ideal, or a high degree of clarity and attack is required, each section will require a microphone and the overall pair will have to work closer, but none should be nearer than a few metres.

There are some concert halls in which most of the reflective surfaces are concentrated behind the orchestra, with a non-reverberant auditorium. This can have the effect of assisting the players who need it least (i.e. the brass and percussion) at the expense of the strings. In these conditions a fairly close microphone technique is usually required.

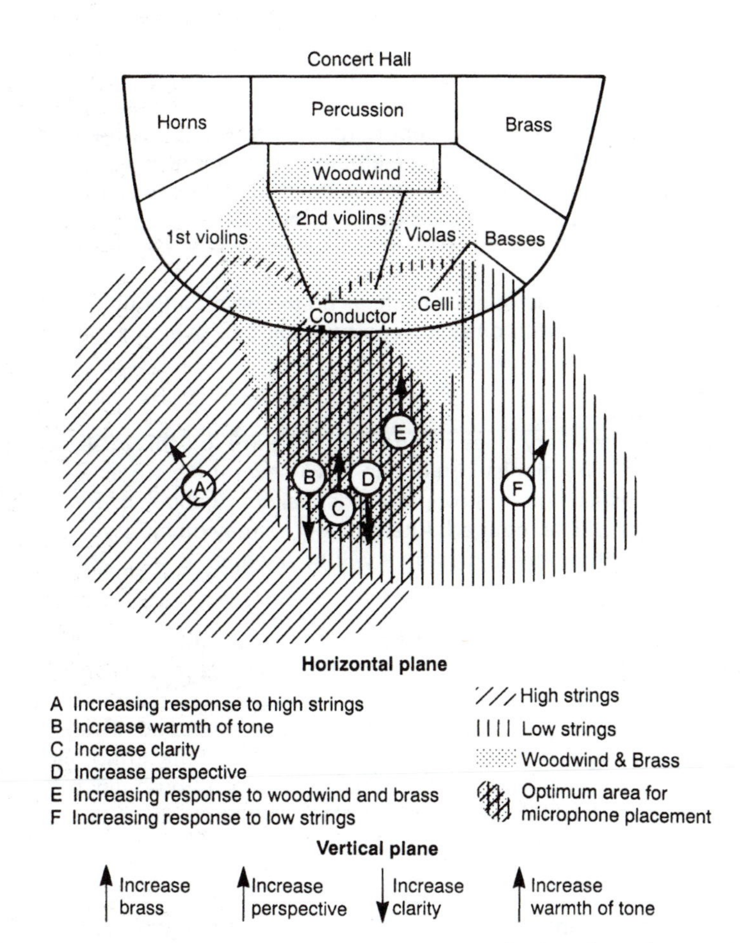

Factors to be considered in positioning a main overall microphone (or stereo pair) for a large orchestra in a concert hall.

Big (Dance) Band

This type of band consists of a brass section (the big bands have 4 trumpets and 4 trombones), a saxophone section (normally 2 altos, 2 tenors and a baritone) and a rhythm section consisting of piano, guitar, bass (string bass or electric bass guitar) and drums.

Rhythm section

The rhythm section is a very important feature of the dance band. In addition to supplying an important element of the sound, which should be reproduced with almost equal volume to the tune, it drives the rest of the band along and gives 'lift' to their performance.

Brass and saxes

The saxes require a microphone each, as do the brass, although it may be possible to manage with one between two and one overall for the brass section.

Rhythm piano

The piano requires very close miking (see page 96) to achieve separation from the band. The pianist could also have several electronic keyboard instruments which play through loudspeakers (cabinets) for which individual mics will be required. Alternatively he may do his own mix with a sub-mixer so that only one cabinet will be used.

Double bass

The bass, whether it be a string bass or electric guitar, should sound almost as loud as the rest of the band. This is simple to achieve with an amplified guitar but much more difficult with a double bass which, when played pizzicato, i.e. plucked, produces a very small sound. It requires a unidirectional microphone within a few centimetres of the strings close to the bridge. Alternatively an omnidirectional microphone can be used softly suspended within the bridge, unless the bass has a transducer pickup, in which case the cabinet will need a mic.

Drums

The drums of a dance band have an enormous volume range. Nevertheless, to be reproduced with proper clarity they require a close microphone technique. See Rock Groups (p. 114).

If the drums are on a rostrum it needs to be very firm and acoustically damped to prevent spurious resonances travelling up the microphone stands and blurring the sound.

Guitar

The guitar is almost invariably electrically amplified and a microphone can be placed in front of the loudspeaker cabinet. See also Rock Groups (p. 114).

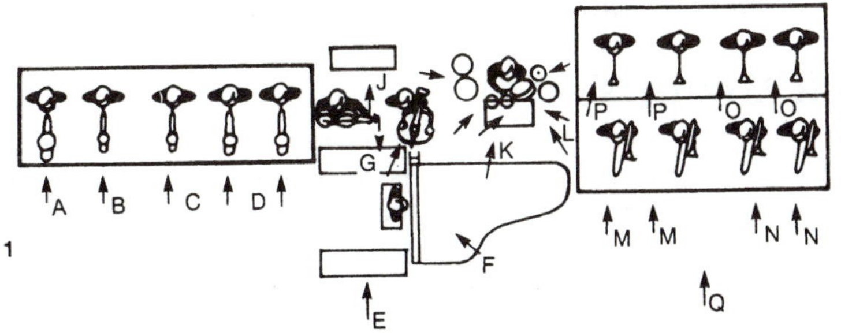

1

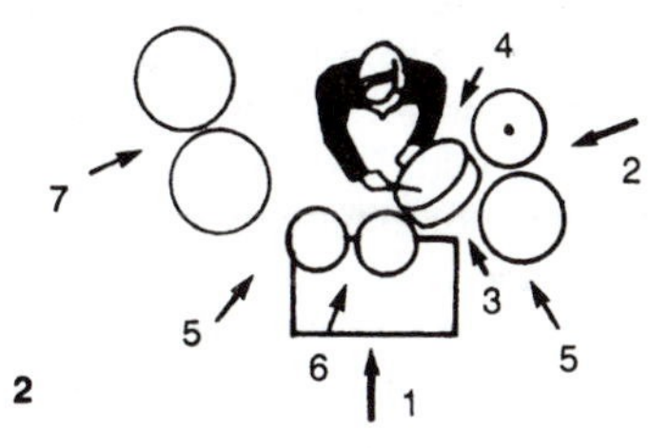

2

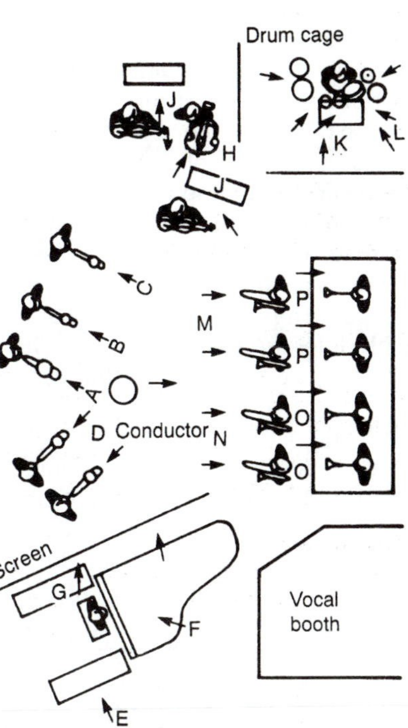

3

1. *Typical layout for TV or stage presentation.* Note the rhythm section in the middle to keep the musicians (who are really too separated for ensemble) in time. The microphones (arrowed) cover A, baritone saxophone (sometimes doubling flute). B and C, alto saxes. D, tenor saxes. E, synthesizer and cabinet. F, piano. G, keyboard synthesizer. H, double bass. J, guitar amplifier. K, bass drum. L, side drums and hi hat. M and N, trombones. O and P, trumpets. Q, overall brass microphone (to give cohesion and a sense of power to the brass sound).

2. *Overhead view of drum kit.* A close microphone technique is required to produce a crisp attack. Short-decay reverb can be added to the kick (bass drum) to give a more powerful effect. The mics will include the following. 1, bass drum. 2, hi-hat (pedal cymbal). 3 and 4, top and bottom snare (side drum). 5, overheads (for cymbals). 6, tom-toms. 7, bongoes. See also Rock Groups (p. 114).

3. Possible layout for a studio recording. The vocalist, whether screened off in the vocal booth or recorded separately, will need to be 'folded back' to the band on loudspeakers and to the conductor and drummer on headphones.

Rock groups rely heavily upon microphones and synthesizers.

Rock Groups

All the instruments in rock groups, with the possible exception of acoustic drums, are amplified through powerful loudspeakers.

Drums

These form the centrepiece of rock music, with a powerful beat that should be prominent in the balance. They can be played acoustically or synthesized or a mixture of both. For good internal separation, the key will typically be covered with two overheads, five or more tom-toms, a top and bottom snare, hi-hat and kick (bass drum). The *snare drum* has a set of springs which can be brought into contact with the lower head, prolonging the note and giving it a 'whiskery' quality which can be 'fattened' by adding reverb with a small room acoustic.

Electronic drums can be used to increase the scope of the acoustic kit. Instead of drums they use pressure pads which trigger synthesizers to produce a variety of sounds such as synthesized drum sounds or musical pitches. The pads are velocity sensitive and can trigger different sounds with different levels of impact. The synth outputs can be mixed on stage and may be fed direct to the recording via a DI box.

Guitars

It is usually best to mic the guitar cabinet (loudspeaker) as this output will include all the player's effects pedals and any (intentional) distortion produced by his amplifier and LS. When guitars are used with cascaded pedal effects, or as triggers for synthesizers, where there is a risk of fingering noises becoming intrusive, it may be necessary to use a noise gate.

Bass guitars may be picked up straight from the instrument with a DI box, but will probably need a lot of bass lift in the EQ, to which some of the cabinet sound can be added.

Keyboards

These are used with synthesizers. Often several keyboards with different synths are used so it is best to rely on the musician's sub-mix by micing his cabinet. Alternatively, if multi-track recording is possible, separate tracks can be made and mixed down later.

Vocals and acoustic instruments

The sound must be fed back to the musicians on stage on monitor loudspeakers. Microphones with clean off-axis cutoff are required to minimize the chance of howl-round. If the vocalist is mobile the microphone/monitor relationship must be checked for all positions.

Vocalist's mics must be robust, preferably with internal shock suspension to stand up to handling, and incorporate windscreens and EQ for very close working (see page 42).

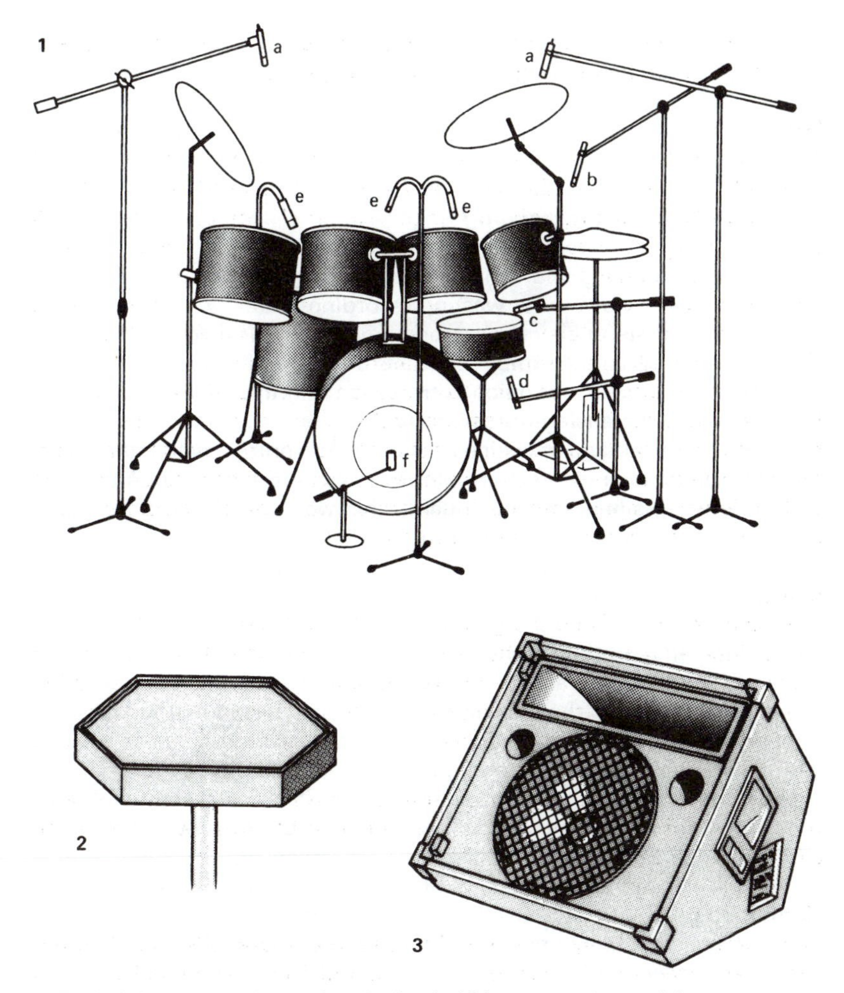

1. Typical drum setup. a, overhead mics. b, hi-hat. c, top snare. d, bottom snare. e, tom-toms. f, kick.

2. Pressure pad to trigger synthesizers.

3. Stage monitor loudspeaker to feed back vocals and acoustic instruments to the musicians.

All microphones must be capable of accepting very high levels without distortion. Their outputs are pan-potted into position in the stereo image.

Multi-track recording is used to save artists' recording time and for synchronizing purposes.

Multi-track Recording

Efficient post-production work and music recording require the use of a multi-track recorder. There are analog multi-track machines that record 4 or 8 tracks on 25.4 mm (1 in) tape or 16, 24 or 36 tracks on 50.8 mm (2 in) tape, or up to 48 tracks on ½ inch tape with a digital recorder (see page 203).

The purpose of multi-tracking

There are three basic reasons for multi-tracking:

1. To enable re-mixing and various forms of manipulation to be accomplished at a later date without wasting recording time.
2. To enable artists to play in accompaniment to their own performance, to increase the apparent number of artists on the final recording, allowing for repeat takes of each addition without spoiling the previous ones and without unduly increasing the noise level.
3. To enable recording machines to be synchronized with other tape recorders, videotape recorders or film machines, using one track to carry a synchronizing signal or time code. In this way sound tracks for video recordings or films can be built up, adding dialogue, effects, music etc. on the various tracks, all synchronized with the picture. They can then eventually be mixed down to form a single synchronized sound track.

Multi-microphone technique

Making a multi-track master recording involves multi-microphone technique. Each section of the orchestra or instrumentalist in a group is given a separate microphone, sometimes more than one to each player (e.g. a drum kit can require up to ten microphones for some types of balance). The microphone outputs are processed in a multi-channel sound console which provides the facility to monitor individual channels (by selection) or produce a 'dummy' mix in which they are mixed into the monitoring system while remaining separated for the recording tracks. Visual monitoring is provided for all channels so that the volume on each of the tracks can be kept within the parameters of the recording system.

At a later stage the various tracks are 'mixed down' to mono, stereo or quadraphonic form. This process can take a considerable time if much editing is involved or if complicated treatment (e.g. various forms of artificial reverberation and frequency response shaping) is required for the individual tracks. It can all be accomplished, however, without retaining the artists.

Multi-track recorder mechanisms

Tape recorders handling wide tapes require powerful motors and well-engineered mechanical systems and electronic tape tension sensors because of the large transfer of weight as the tape spools from one reel to the other.

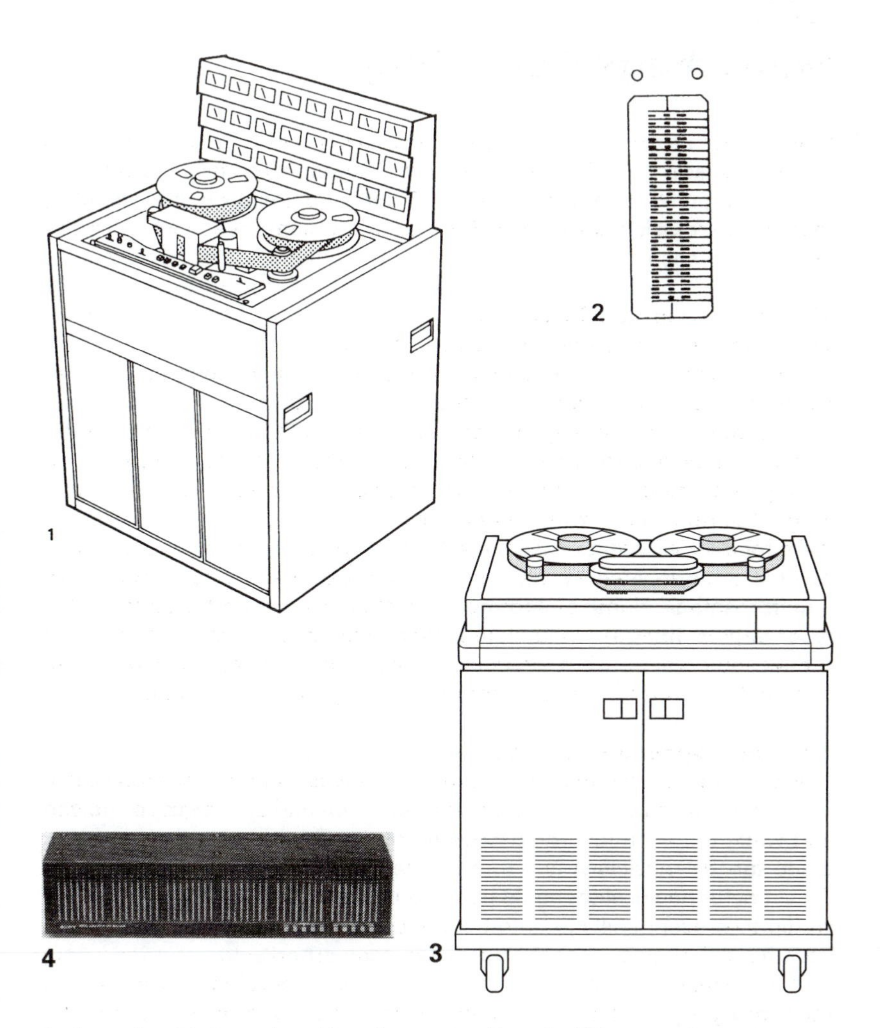

1. An analog 24-channel multi-track recorder. Note the VU meters for each channel. In order that the tracks can be recorded at different times but in sync it is necessary to employ a complicated system of head switching. Individual erase heads are provided for each track.
Although separate record and replay heads are provided, arrangements have to be made to use the record head to playback previously recorded tracks while further tracks are being recorded, to enable the artist to synchronize with the previous recording. (The repro head does not synchronize with the record head.)

2. Typical multi-track record head. Great care must be taken in setting the azimuth, otherwise serious phase differences can exist between the most separated tracks. When selecting tracks for recording it is advisable to place any sources that are likely to have overlapping sound (e.g. two microphones in close proximity) on adjacent tracks. Otherwise phase differences could cause undesirable effects.

3. A 48-track digital recorder. Note the 48 volume indicators (Courtesy Sony).

4. A 48 channel remote metering unit.

Rock groups can benefit greatly from multi-track recording.

Recording Rock Groups (1)

Multi-track recording

Multi-track recording opens up an enormous range of possibilities for rock groups. Digital recording makes it possible to increase the apparent number of musicians, to correct errors and improve each element recorded on each track, without affecting the others. It is possible to start with a simple phrase, or note sequence, and build it up by multi-track recording, manipulating each track independently before the final mix-down to the master recording.

Using a computer-driven synth coupled with MIDI it is possible to alter the tempo of a recording without affecting the pitch, so that a difficult sequence can be recorded more slowly and brought up to speed once the notes have been sorted out.

Above all, multi-tracking opens up great possibilities for artistic expression, by the ability to experiment with various types of embellishment for each element of the sound before finalizing the mix.

For a sophisticated setup, working in the digital domain, the basic requirements are a mixer and multi-track recorder, possibly hard disk or tape, with anything up to 48 tracks: MIDI to enable the outputs of several synths and other sources to be linked together and synchronized, effects units including digital reverb and frequency response shaping, in addition to that incorporated in the mixer, and microphones for pick-up of studio sound from acoustic instruments and vocals etc.

The vocal can be recorded separately from the band (for acoustic separation) although it is usually advisable for the vocalist to record the part while the band recording is taking place. This can then be used as a guide for the musicians and to help the vocalist when making the final vocal track. The vocalist should be in a cubicle or a screened-off part of the studio to prevent spillage onto the band mics. Each band member must have headphones to hear the vocal and each other.

Bass drum

When making the multi-track recording it is usually best to start by concentrating on the drums which, after the vocal, form the centre piece of the balance. The aim is to produce a clean, crisp sound with a clear 'leading edge' and just enough reverberation or 'thickening' to give the effect of power without 'ringing'. It is usually necessary to damp the bass drum by removing the rear head (skin) or by cutting a hole in it through which absorbent material can be introduced and the microphone can be inserted or given a clear view of the front skin.

Rock groups can provide many opportunities for experimentation.

Recording Rock Groups (2)

The snare drum

In the case of the snare drum it may be necessary to mic both the top and bottom heads to get a good balance between the solid sound of the top head and the 'whiskery' sound of the snares (the springs stretched across the bottom skin).

However, when micing both sides of a double-sided drum the two sides will be out of phase. High pressure on one side of the skin coincides with low pressure on the other, so it will be necessary to reverse the phase of one microphone, usually the lower one, to prevent cancellation of their outputs.

The top head of the snare drum can usually do with a small amount of damping, the time-honoured method being to place a folded duster on part of the skin. Some drums have a built-in damper. When not in use there is always a possibility of the snares (springs) rattling due to excitation from the pressure waves from other instruments (particularly the bass drum). When not in use the spring can usually be separated from the skin with the use of a lever on the side of the drum.

Microphones

Microphones used around drum kits, and especially if placed inside the bass drum, must be capable of withstanding very high pressures. The cymbals and toms require overhead mics and the hi-hat a mic of its own. The vocal mic can be shielded by a mesh screen in the studio situation, but those used on location must be robust and capable of working very close, if not actually touching the vocalist's mouth.

Mix-down

Instruments, whether picked up by direct injection or by micing their cabinet, can be recorded separately, the amount and length of reverb and possibly EQ being arrived at by experiment before the final mix-down is accomplished. However, it is advisable to make a 'dummy' mix while the individual sources are being worked on to ensure that there are no clashes or unwanted effects when the final mix-down is made. It may also be necessary to establish the relative position of each source in the stereo image by adjusting the pan controls on each channel.

Equalization

While it is usually best to apply EQ to each channel individually, to make it possible to assess the effect on each instrument, it could be worthwhile to apply a certain amount to the final mix, to give the sound a particular character.

Four-track Recording

Budget recording
The simplest way to get started with multi-track recording (for making reasonable quality demonstration tapes or if funds are tight) is to use four-track cassettes. Special equipment is available which combines the sound mixing console and cassette recorder in one unit. In the usual format for recording on cassette, two stereo tracks are recorded one way on the tape and by turning the cassette over, to double the playing time, two tracks are recorded on the other side in the opposite direction. However, in this case all four tracks are recorded in the same direction on one side, providing full four-track recording. To improve the quality, the tape speed can be double (9.5 cm/sec instead of the normal 4.8 cm/sec) so that a C90 cassette will only provide about 22.5 minutes of playing time. However even the best-quality cassettes are cheap and can be used many times. Dolby C or DBX is used to improve the signal/noise ratio.

Four-track recording rock groups
When making a multi-track of a rock group the first item to be recorded is usually the drum track, as this is necessary to keep the other instruments in sync. The start of the drum track can be preceded by a count-in in tempo, beating two bars on a closed hi-hat or sidedrum rimshot, leaving out the last beat so that when the final mix is made there is a gap in which to start the recording. Where more than four tracks are required several tracks can be pre-mixed together and 'bounced' on to one track. For instance, while listening to the drum track recorded on track 1 another track, e.g. the bass, can be recorded on track 2. Another instrument can be recorded on track 3 and then tracks 1, 2 and 3 can be mixed in the correct proportions and bounced onto channel 4. Once a good balance has been obtained tracks 1, 2 and 3 can be erased and these tracks used to record other instruments or vocals and their tracks mixed and bounced as necessary. In this way much of the effect of a much larger mixer can be obtained from the four channels. Most four-track combined recorders incorporate the facilities normally available in small mixers including the ability to 'drop in' to the record mode on individual tracks, possibly to replace a wrong note or other blemish.

Mastering
Finally the four tracks are mixed down to stereo while adding reverb, and EQ if required.

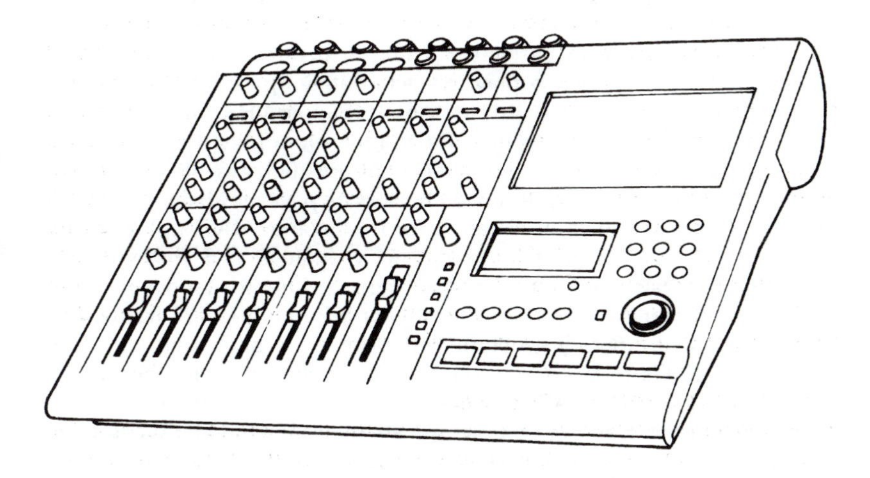

The TASCAM 464 four-track 'PORTASTUDIO' shown here is an example of a complete four-track recording unit.

The mixer section has twelve input channels: Standard mic/line input channels 1 to 4, stereo line input channels 5 and 6 and 7 and 8, stereo sub input channels 9 and 10 and 11 and 12. All twelve inputs can be fed to the stereo bus and mixed with three tape tracks while track 4 is used for synchronization. The Buss Direct System makes it possible to mix twelve synchronized MIDI sources with three tape tracks, a total of fifteen 'tracks'. Three-band EQ with sweepable midrange frequency is provided on the mono inputs for response shaping.

The recorder section has selectable tape speeds: 9.5 cm/sec ($3\frac{3}{4}$ in/s) for quality and 4.8 cm/sec ($1\frac{7}{8}$ in/s) for compatability with standard cassettes. Built-in DBX noise reduction, pitch control ±12%, direct record mode, allowing simultaneous four-track recording, auto punch in/out capability with rehearsal mode. Sync IN and OUT jacks allow track 4 to be connected to an external synchronizer.

This type of equipment can provide quite adequate facilities for the small studio with a tight budget or someone starting to gain experience in multi-track recording.

ADAT

ADAT

ADAT is a digital multi-track sound recording system developed by the Alesis Corporation. It works on a similar principle to R-DAT, i.e. it is a rotary-head helical scan cassette system. The rotating drum has four heads (two record and two playback alternately spaced at 90° to each other). The recording medium is standard Super-VHS tape cartridges and tape transport is as used in VCRs and therefore fully developed. The tape cassettes are readily available and inexpensive, S-VHS being chosen for its high density and coercivity (data packing ability). The tape runs at 3¾ in/s (about three times faster than standard VHS play speed) and offers up to 40 minutes of audio on a single S-VHS cassette.

The recorder has eight tracks and is designed either to stand alone or to link up with other ADAT machines, providing from 16 to 128 tracks. The ADAT built-in proprietary synchronization interface allows an unlimited number of ADATs to be locked together in synchronization. The remote control enables several ADAT decks automatically to start recording as each one nears the end of its tape and seamlessly plays the tapes back. The remote control also reads SMPTE time codes and generates MIDI time code and MIDI clocks to provide all the timing reference needed to drive a complete MIDI sequencing system. A dual system is employed to synchronize the position of the head to the movement of the tape. The physical positioning is controlled by a longitudinal control track and a high resolution timing reference provided by timing markers recorded at the end of every helical scan as part of the initial formatting of each tape.

ADAT delivers 16-bit linear audio performance using built-in 64-times oversampling delta sigma A/D converters and the professional 48 kHz sampling rate, user variable from 40.36 kHz to 50 kHz. Separate converters are dedicated to the inputs and outputs of each individual track so that multiplexing is not required.

The long tapes and relatively large recording medium (compared with some types designed for portable use) result in rather slow rewind times, although the occasions when it is required to spool from end to end are likely to be limited.

The CD-ROM and CD Plus

CD-ROM (compact disk read only memory) was developed mainly for use with computers. It is a data storage system with a capacity of about 600 million bytes, the equivalent of 200,000 written pages of text. It can also store pictures and sound.

Physical characteristics
The CD-ROM disk looks very similar to a music CD, but there are differences in the way the information is organized. The CD-ROM requires a more powerful error correction system than the music CD, because it is important that no information is lost which could be vital, but might go unnoticed or sufficiently corrected in a music CD.

Data is not formatted in the same manner for the two media.

CD-ROM players
In an audio player the D/A converts the full output to analog form, but the CD-ROM output remains digital and is therefore suitable for application to a computer.

CD-ROM drives must have very fast seek times to keep up with the information demands of a computer, whereas audio disks can be allowed a few seconds delay between selecting the item and the start of play.

Drive speed
The original CD-ROM drives were single speed, operating at about the same speed as floppy disks. Later versions offered double or quadruple speed, with corresponding increase in data transfer rate. These are standard speeds at present, but even faster 'quad plus' (4.4 times) and 'six speed' are becoming available. The majority of CD-ROM drives have the same drawer system as hi-fi players for inserting the disc, but some players use a caddy system where the CD is inserted into a case about the size of a normal CD box, for protection against dust before insertion into the drive.

CD Plus
Sony and Philips have released a specification for a new 'enhanced music' CD which combines normal CD audio tracks with CD-ROM data, which they call CD Plus. Personal computers equipped with CD-ROM drives will be able to show still pictures, titles and lyrics as well as playing the music. The audio tracks are playable on existing audio players, so that prerecorded discs need only be issued in the one format.

MIDI

MIDI stands for musical instrument digital interface. In its simplest form it is a standardized method of interconnecting electronic musical instruments and computers.

Linking together synthesizers, drum machines, sequencers and keyboards can require a large number of interconnecting cables to cater for all the instructions that have to be passed down the line. MIDI is a purely digital system that uses serial interfacing so that the commands can be serialized in digital code and fed down a single twin cable one bit at a time, the equipment only having to recognize level 0 or 1. Twin screened cable is used, not so much to prevent the cable picking up interference and hum, as to prevent it radiating radio waves which could be picked up by the equipment due to the high frequencies involved.

Opto-isolators

MIDI links incorporate an opto-isolator, consisting of a photo-transistor and LCD facing each other in an opaque capsule. The photo-transistor is switched on by the light from the LCD so that a switching action is transmitted through the device without an electrical connection. This is important when connecting units that are not earthed, where a substantial potential difference could build up between them.

Control system

The command data is sent down the cable one bit at a time, forming bytes of eight bits, preceded by a start bit and followed by a stop bit. The start bit is also used to synchronize the transmitting and receiving systems, which are normally kept in sync by quartz oscillators. These bytes form numbers denoting instructions, one to three bytes being required according to the complexity of the data. They are sent around the system at a rate of over 3,000 bytes per second. In addition to the 'in' and 'out' sockets MIDI controllers and instruments can have a socket labelled 'thru'. These are extra outputs fed through a buffer amplifier to enable several instruments to be linked together without overloading the system.

The combination of MIDI-equipped instruments and a computer can open up great possibilities, especially when arranged as a multi-track recorder. For instance music can be recorded one track at a time, and stored in the computer, if necessary at slow speed, and played back at the correct speed, building up to a complex musical arrangement. Notes can be adjusted for length, correct notes substituted for wrong ones and instruments changed, even on the same track, if it is found to improve the result.

MIDI has sixteen channels available to perform different functions, apparently simultaneously, by directing information to one particular device in a system. Instruments receiving MIDI information specific to a

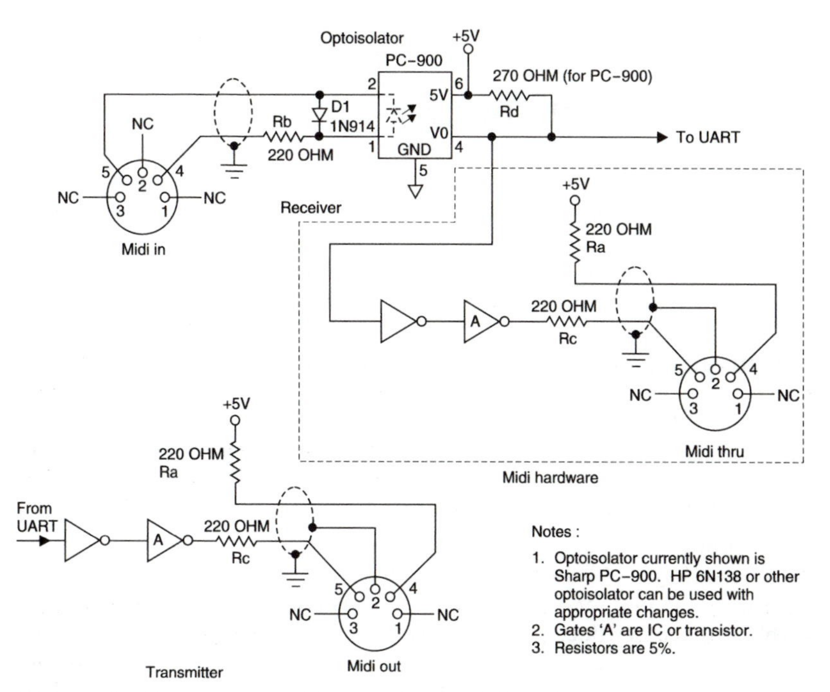

MIDI standard hardware specification. Note the 'in' and 'out' connects are both sockets. Connecting cables require male connectors at both ends, the equipment is connected to pins 4 and 5 of the 5-pin DIN connectors, pin 2 being connected to the screen at one end only.

particular channel can be assigned to one of four different modes which govern the way each piece of equipment handles the MIDI channels.

MIDI modes

The standard MIDI operating modes are numbered 1 to 4 and named as follows:

Mode 1 (Omni on/poly). The default mode that is automatically adopted on switch on. It responds to notes on all channels polyphonically, i.e. it can handle several notes at once.

Mode 2 (Omni on/mono). Responds to all notes on all channels monophonically. (Intended for use with monophonic instruments.)

Mode 3 (Omni off/poly). Responds to notes on one channel polyphonically, ignores all note information that is not on a specific channel. It provides polyphonic operation provided that the transmitting device and receiving instrument are set to the same channel.

Mode 4 (Omni off/mono). Responds to notes on one channel monophonically. Each voice of an instrument is assigned to a separate MIDI channel. With a sixteen-channel, multi-timbral instrument it would be possible to have sixteen different sounds each sequenced independently.

Sound can have an enormous volume range. The trick is to preserve the psychological effect without exceeding the technical parameters.

Signal Processing

Programme makers naturally want their programmes to be heard and that means all of it, the quiet bits as well as the loud. Most domestic equipment, especially television sets, is not capable of reproducing the range of sound encountered in real life, and the viewers would not put up with such volume in the domestic situation anyway.

Levelling and compression

Levelling is the long-term control of signal level to maintain a relatively constant output level without lessening the impact of short-term variations which have been 'tamed' by the use of *compression. Process balance* determines the ratio between levelling and compression.

The 'ducker'

This is a form of compressor in which the output of one signal is controlled by another. It can be used for 'voice-overs' to reduce (duck) the level of music or effects when a narrator or DJ speaks.

The noise gate

When compression is applied to a signal it brings up the gain when the level is low and that includes the noise, which will be obvious in gaps in the programme. To overcome this a noise gate can be used. It can consist of a simple electronic switch which cuts off the output in the absence of signal and restores it when the minimum acceptable level is reached. Unfortunately in many circumstances the sudden switching of noise only tends to draw attention to itself. A more elegant method of noise gating is the *expander.* This is virtually the opposite of the compressor. It has a gain greater than unity up to the threshold point so that it effectively increases the distinction between wanted and unwanted sound.

De-essers

Simple compressors operate on the most powerful component in the frequency range of the programme. In the case of speech it is the lower middle frequencies in most words that drive the compressor, reducing the gain, which can be restored in time to accentuate the weaker consonants (particularly 's' sounds at the ends of words), resulting in an unpleasant sibilance. Many compressors employ de-essing circuits which introduce pre-emphasis into the side chain to equalize the controlling effect.

Multi-band peak limiting

One problem in the use of limiters is the effect of 'pumping' when strong signals of one frequency take control of lower-level programme. For example, a bass drum beat could cause the general sound of a band to 'pump' up and down. This can be overcome by dividing the limiting action

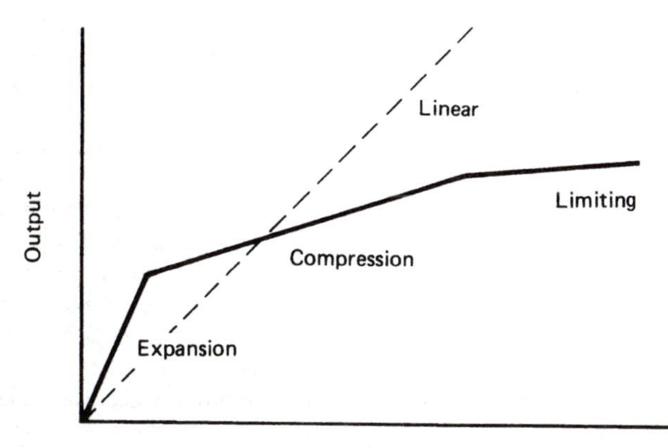

A composite compressor, incorporating an 'expander' to enlarge the distinction between wanted and unwanted sound (a form of noise gate), a compressor to 'tame' widely varying material and a limiter to catch overshoot.

into frequency bands (usually low, middle and high) with separate threshold and slope controls for each. This can also help to solve the de-essing problem.

Companding
This is the process of improving the signal/noise ratio by compressing the volume range of a signal before a process (such as recording or line transmission) and expanding it afterwards. The problem is to make the two functions track accurately throughout the range. (See p. 128.)

Dynamic Noise Reduction: The Dolby A System

Modern recording techniques usually involve the re-recording (dubbing) of programme material from one medium to another, in some cases many times over. This can be for the purpose of 'reducing' a multi-track recording to twin-track stereo, for editing or for transfer to disc etc. Unless digital recording is used, each recording will add its quota of noise to the final result.

Companding

In order to improve the signal-to-noise ratio a system of 'companding' can be used, i.e. the volume range of the signal is compressed to raise the average level before recording and then expanded on reproduction, depressing the quieter elements of the programme (and with it the noise).

Unfortunately there are two main problems in designing a companding system that does not cause more trouble than it cures. One is the difficulty of making the compression and expansion systems 'track' perfectly over the whole frequency and volume range. The other is preventing the background noise level going up and down with the signal, which only makes it more obvious. The *Dolby A system* overcomes this problem in the following ways.

The signal is provided with two parallel paths, one via a linear amplifier and the other via a differential network, the output of which is added to the 'straight through' signal when recording and subtracted when reproducing. The differential network divides the frequency spectrum into four bands and treats each one separately so that the action is applied only where it is needed. With music, for instance, most of the sound energy tends to be concentrated in the lower middle band, whereas tape hiss is largely high and too far removed in frequency for 'masking' to take effect. This is the effect whereby the ear tends to be deaf to sounds in the presence of, and for a short time after, hearing louder sounds of similar frequency. The Dolby system exploits this effect by restricting the action to narrow frequency bands.

The system is adjusted so that with a low level input (below −40 dB) the output is 10 dB higher than the input for frequencies up to about 5 kHz, above which it increases progressively to 15 dB gain at 15 kHz. As the input signal rises above +40 dB, it is progressively less affected. At high levels (where the possible ill effects of companding are most obvious), the compander has least effect and is virtually bypassed. The Dolby A system is normally used only as an intermediary process, the final product having a straightforward response. This is in contrast to the Dolby B process, where the recording is sold with the modified characteristic incorporated on the assumption that the complementary characteristic will be applied in the reproducing apparatus.

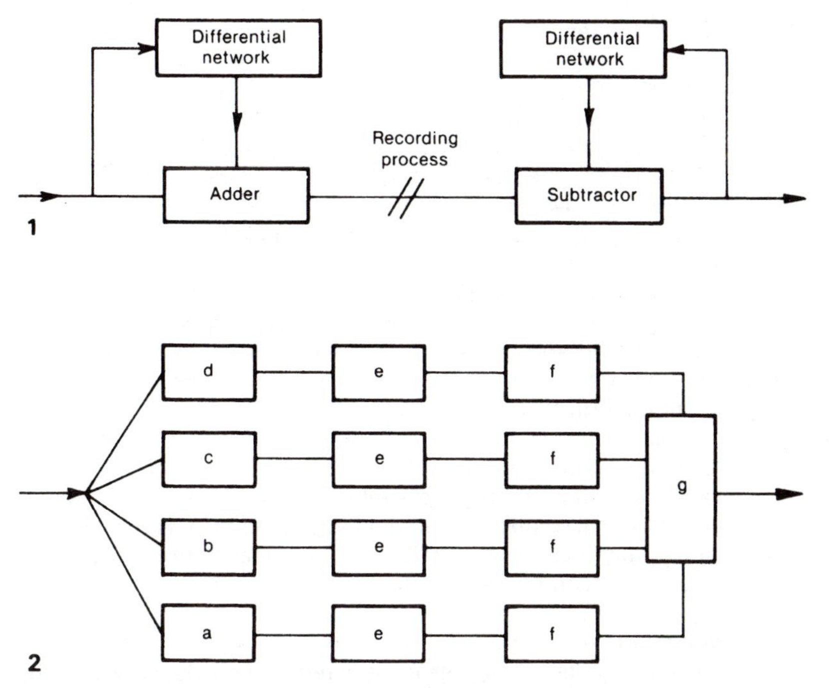

1. Companding the recording process. The output of the differential network is added for recording and then subtracted when replaying.

2. Schematic detail of differential network: a, 80 Hz low-pass filter to deal with hum and rumble; b, 80–3000 Hz band-pass filter to deal with mid-band noise and print-through; c, 3000 Hz high-pass filter to suppress hiss and modulation noise; d, 9000 Hz high-pass filter to suppress hiss and modulation noise; e, linear limiters; f, non-linear limiters; g, adder.

The Dolby B noise reduction system is used in high-quality domestic cassette recorders.

The Dolby B Noise Reduction System

A simpler system than the Dolby A process is the Dolby B system, which is used mainly in the domestic tape cassette market to reduce the high background level of tape hiss which is inherent in this type of recording due to the low tape speed and narrow track. With Dolby B the tapes are recorded and marketed with the Dolby characteristic on the assumption that they will be reproduced by equipment with the complementary characteristic.

As in the Dolby A system, there is a main programme path and a side chain. The side chain incorporates a compressor which is preceded by a high-pass variable filter covering frequencies from about 500 Hz upwards. In the record mode, signals below a given threshold level are boosted by the compressor and added to the side chain. This increase in level is applied progressively by the variable filter from 500 Hz up to 10 dB at 10 kHz.

Thus low-level high-frequency signals are recorded up to 10 dB higher than the original. An overshoot suppressor (diode clipper) is provided, following the compressor, to prevent high-level transients which are faster than the time constant of the compressor from being added to the output. Any resultant distortion is masked by the high-level signal and tends to cancel on replay.

The Dolby B decoder, which is fitted to most high-quality cassette machines, is identical to the coder except that the side chain is fed with a phase-inverted signal so that the output of the compressor is added to the main chain in antiphase and is effectively subtracted from it, thereby producing an exact complement of the coding process.

Noise reduction

As the low-level high-frequency response is reduced on playback so is the tape hiss and system noise that has been introduced by the recording, giving an improvement in signal-to-noise ratio of up to 10 dB.

The difference between the Dolby system and the simple application of pre-emphasis to the recording and de-emphasis on playback (which is applied anyway) is that the Dolby B characteristic affects only low-level signals.

Compatibility

Dolby B encoded material can be played on equipment without the Dolby function if the high-frequency response is reduced to compensate for the Dolby characteristic, but this does result in a lack of high frequency in the louder passages of the music.

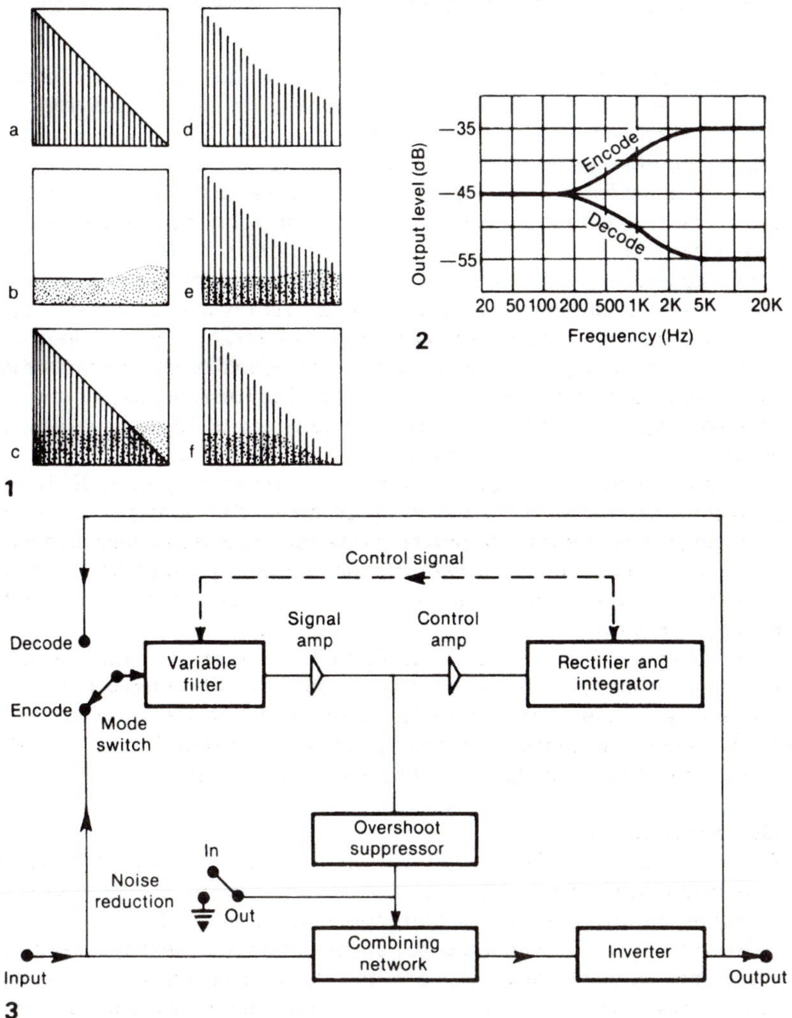

1. (a) A representation of the distribution of energy in average programme material. The energy tends to decrease as frequency increases. (b) Tape noise. (c) In normal reproduction low-level high-frequency signals are masked by tape hiss. (d) Dolby B encoded signal with compression applied to boost low-level high-frequency signals. (e) When the tape noise is added the boosted high-frequency signal is still above the noise level. (f) Decoding reduces the low-level high-frequency response to normal, and with it the noise.

2. Complementary response curves of a Dolby processor with low-level input.

3. Block schematic diagram of a Dolby B encoder/decoder.

An extension of the Dolby B principle can produce even more effective noise suppression.

The Dolby C Noise Reduction System

Dolby B (page 130) is a noise-reduction process in which 'sliding-band' compression is applied in the making of a recording, on the assumption that complementary expansion will be used in reproduction. It provides up to 10 dB of noise reduction.

Dolby C is an extension of the Dolby B principle. It offers up to 20 dB of noise reduction with improved protection from recording distortions and unwanted side effects. It employs two compressors in series in the recording process and two complementary expanders in reproduction, the first stage operating at levels comparable with B-type systems (which can make use of similar circuitry) and the second sensitive to signals 20 dB lower in level. The result of adding the two companders is to increase the noise reduction to 20 dB in the critical 2 to 10 kHz area. Whereas the type B system begins to take effect in the 300 Hz region and increases its action to a maximum of 10 dB at 4 kHz and above, Dolby C begins to take effect in the 100 Hz region, providing 15 dB of noise reduction around 400 Hz; it thereby reduces the 'midband modulation effect' in which middle frequencies become modulated by high-level high-frequency signals, owing to a mistrack between the encoded and decoded signals.

Spectral skewing

The problem of midband modulation and other unpleasant effects, due to the unpredictable response of cassette recorders at the extremes of their frequency range, is overcome by a process called 'spectral skewing'. This consists essentially of a steep roll-off in the input to the high-level stage of the compressor to make it insensitive to signals above 10 kHz. A complementary boost above 10 kHz is applied at the output of the final expander.

Antisaturation

To reduce the risk of overload and tape saturation at high frequencies, with consequent risk of compander mistracking, a shelving network is placed in the circuit which affects only the high-level signals. At low levels most of the signal passes through the side chain. A complementary network is provided in the expansion mode to maintain a flat response.

Owing to the two-stage design of the Dolby C system it is relatively simple to design equipment with the facility to switch between Dolby B and C.

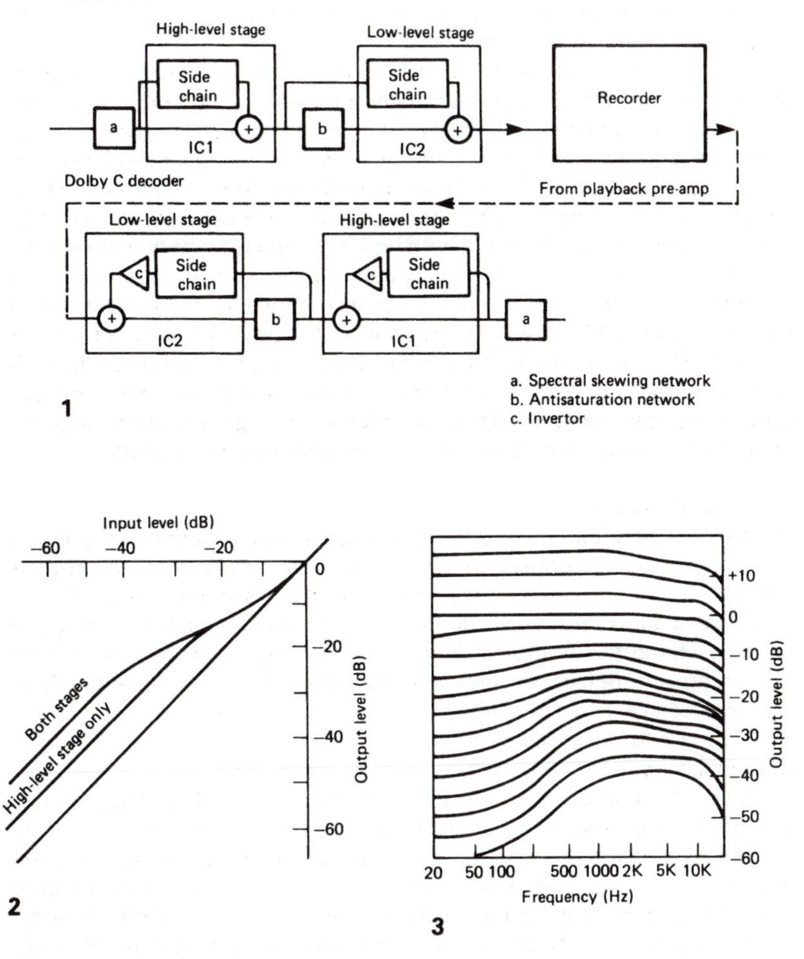

1. Block schematic diagram of a Dolby C encoder/decoder.

2. Input–output transfer characteristic of C-type compressor at 1 kHz, showing the effect of high-level stage on its own, and both stages in series.

3. Dolby C compression characteristics.

Dolby SR and S Noise Reduction Systems

Dolby SR (spectral recording)

The Dolby SR noise reduction system is intended for use on the input and output of professional tape recorders, like the Dolby A system, for which it can be a replacement. As with all Dolby systems it is fully complementary, i.e. the decode process is a mirror image of the encode processing. It is, in fact, an extension of the Dolby B and C systems, but whereas these are intended to deal mainly with slow-speed cassettes, in which the noise problem concerns mainly the high frequencies, the SR system is designed for very high quality recording with high tape speeds where mid- and low-frequency noise and print-through can be a problem.

Whereas Dolby C can be considered as two Dolby B-type circuits working at different levels, each stage feeding into the next, so that low-level HF is amplified twice (thereby removing it from the noise), Dolby SR takes the process a stage further. Three upward-sliding-band HF filter circuits are used, each feeding the next with a gain of 8 dB per stage. This results in a possible 24 dB of noise reduction for frequencies between 800 Hz and 6 kHz. Two downward-sliding-band circuits are also incorporated to give 16 dB of noise reduction below 800 Hz. The sliding-band techniques ensure that processing is applied only where it is needed and thus makes use of the masking effect of our hearing (see Dolby A, p. 128).

Antisaturation and 'spectral skewing' are applied as in Dolby C (see p. 132). The antisaturation network is a shelving filter that reduces the recording level of frequencies at which the tape cannot handle high levels. It is bypassed at low signal levels. Spectral skewing is applied at both ends of the scale to desensitize the system to response errors at the extreme frequencies due to 'head bumps' and HF losses etc.

Dolby S type NR

Dolby S type NR, like the B and C types, is a system in which cassette recordings are marketed encoded with the S-type characteristic and the user equipment applies the complementary decoding. It is virtually a simplified version of Dolby SR optimized for cassette recording. It provides two staggered stages of both fixed and sliding NR at high frequencies (instead of the three for SR) and only a single-stage fixed NR band at low frequencies. Spectral skewing and anti-saturation are provided at low frequencies as well as high to reduce equalization distortion. A 'modulation control' circuit optimizes the dynamic range to suit the music spectrum.

High standards apply to Dolby S licensed cassette machines; these include wider frequency range, low wow and flutter, improved headroom and a standard for head azimuth.

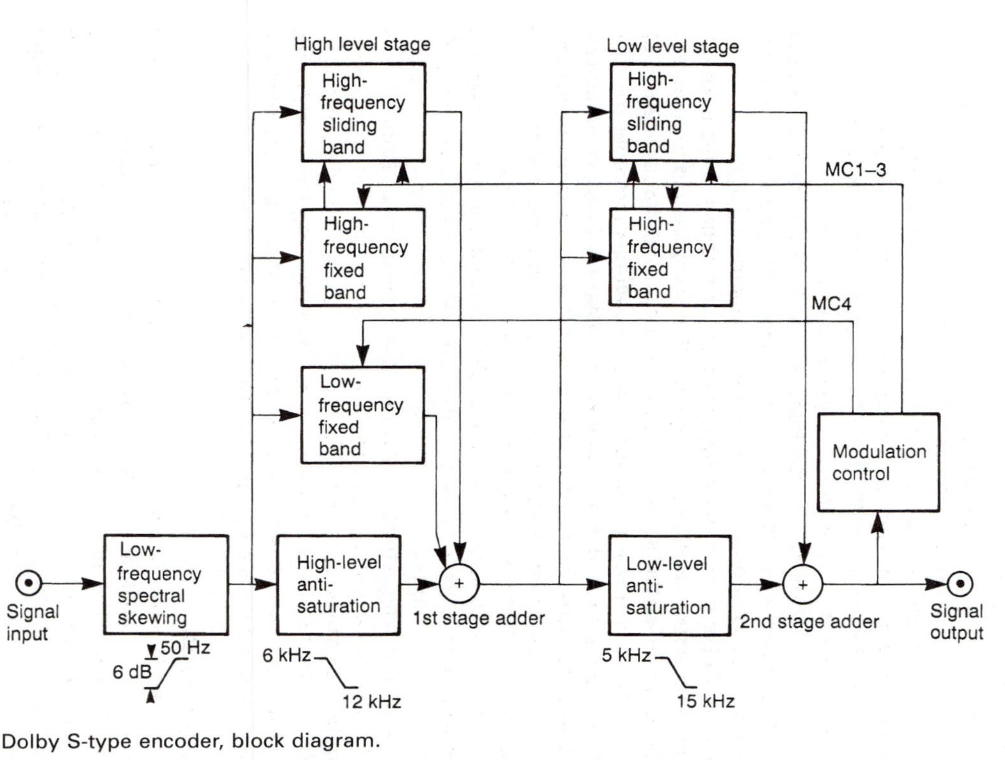

Dolby S-type encoder, block diagram.

As with C-type NR, an S-type encoder has two staggered-action compressors, each having a passive main path which is summed with an active side chain, and each of which operates over a different signal level range.

DBX and DNL noise suppression systems rely on the premise that high-frequency energy tends to increase disproportionately with a general increase in volume.

DBX, DNL and BTSC Noise Reduction Systems

DBX

DBX noise suppression, like Dolby A, is a complementary system which employs a wide range (2:1) compression ratio to encode and decode. Problems of tracking the compression and expansion are eased by the straightforward ratio and the fact that the level sensing is done on an RMS basis, i.e. controlled by the total power of the signal, regardless of the phase relationships of the various components.

The DBX system exploits the fact that, in most programme material, the bulk of the power is in the low frequencies, high power in the high frequencies occurring only when the general volume is large.

The signal fed to the compressor is heavily pre-emphasized to increase the overall power on recording. It is similarly de-emphasized to restore it to normal (while at the same time reducing the high-frequency noise) when it is decoded. To prevent the recording being overloaded by any powerful pre-emphasized high-frequency signals, a similar pre-emphasis is applied to the side chain of the compressor so that, for high levels, the high-frequency recording level decreases as the frequency increases and increases as the high-frequency level decreases.

DBX also takes into account the masking effect of our ears, which makes us relatively insensitive to noise (in this case high-frequency noise) in the presence of loud sound of a similar frequency. The use of DBX can result in an improvement in high-frequency signal-to-noise ratio of up to 30 dB.

Philips DNL system

The Philips DNL system operates on replay only, and therefore must reduce the fidelity of the recording to some extent. As most musical instruments have fundamental frequencies below 4.5 kHz and when played softly produce few harmonics, it is considered that a high-pass filter operating above this frequency has little effect on quiet material.

The system is basically a dynamic treble cut filter operating above 4.5 kHz on low-amplitude signals. When the signal exceeds the pre-determined level, the filter is by-passed thus leaving the harmonics unaffected.

The system is claimed to give an improvement in signal-to-noise level of 10 dB at 6 kHz, 20 dB at 10 kHz.

BTSC noise reduction

As in Dolby and DBX, the BTSC system exploits the masking effect of our ears to sounds of similar frequency. Separate low/middle and HF compression is provided. The LF band operates up about 2 kHz and the HF band rises from 500 Hz to a peak at 10 kHz. There is also protection against overmodulation.

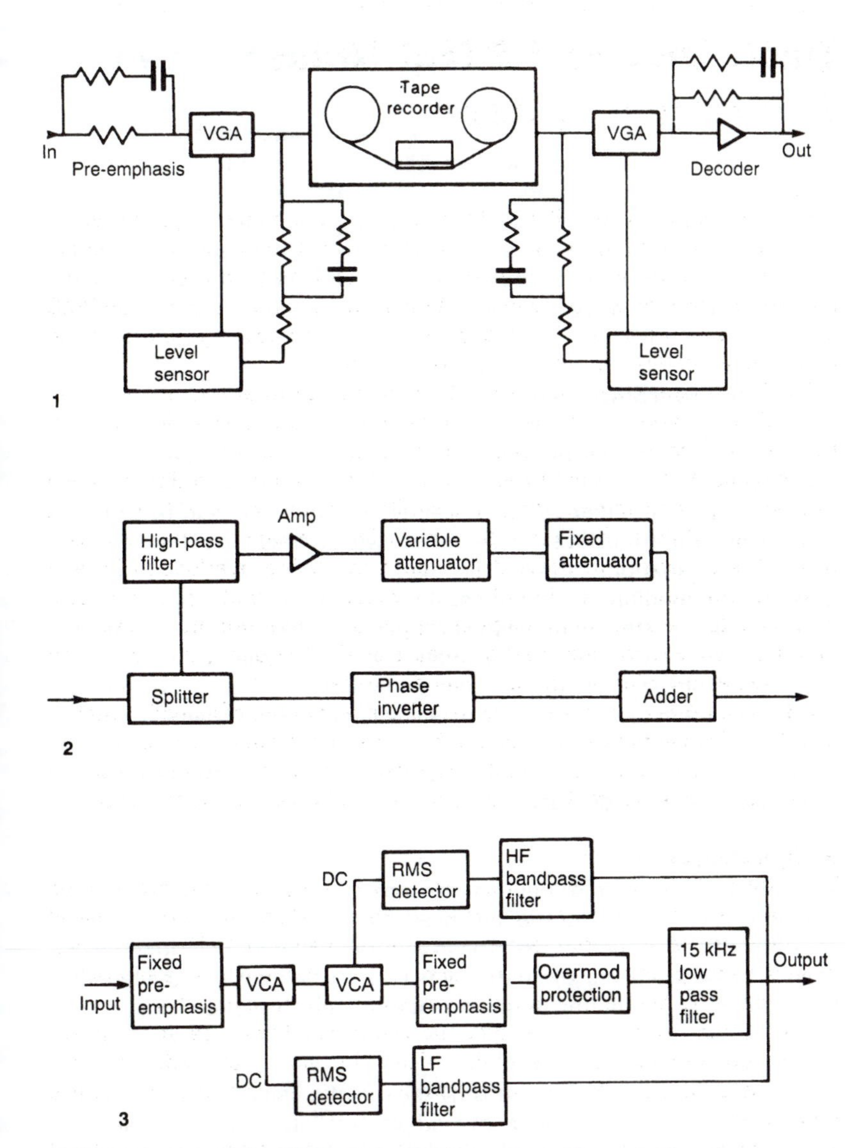

1. Block diagram of the DBX noise reduction system. The variable gain amplifiers have a compression/expansion ratio of 2:1. Due to pre-emphasis, the increase is applied progressively to the high frequencies.

2. Block diagram of the Philips Dynamic Noise Limiter. This works on replay only and reduces the high-frequency response when the high-frequency component of the signal is low.

3. Block diagram of the BTSC noise reduction system.

Magnetic recordings can provide convenience, flexibility and high quality.

Basic Magnetic Recording System

The majority of recording involves the use of magnetic tape. This is because of its convenience and flexibility, particularly from the point of view of editing. Magnetic tape can be played back immediately after it has been recorded and, in the case of professional machines, monitored as it is being recorded. It is possible to erase and re-use the tape; several tracks can be recorded simultaneously and in synchronism and, above all, magnetic tape is capable of producing recordings of the highest standards.

The basic recording system

Tape recorders are made in a wide variety of types and standards of complexity and quality but the basic system consists in essence of the following:

1. A recording amplifier which processes the signal prior to recording.
2. A recording transducer (tape head) which translates the electrical signals into variations of magnetic force.
3. A storage medium, i.e. the tape.
4. A transport system to move the tape past the head at a steady speed so as to equate the variations of magnetic force with time.
5. A reproducing transducer (the original recording or a separate reproducing head) to convert the stored magnetic variations back to electric signals.
6. A reproducing amplifier to process the signal from the reproducing head to provide the output.

Tape transport

The standard direction of travel of the tape is from left to right as viewed from opposite the front of the tape heads. In some machines it is possible to reverse the direction of the tape in order to allow for very long record/replay times by playing alternate tracks (narrow bands along the length of the tape on which the recording is made) in opposite directions.

Mechanical systems

The tape is pulled past the heads by means of a capstan spindle against which it is pressed by a rubber-covered pinch roller. In some machines the tape storage and take-up spools are driven by the same electric motor as the capstan through a system of pulleys. In high-quality machines separate motors are used for each function.

Most recording machines can erase previously recorded material from the tape prior to recording. This is achieved by means of an *erase* head over which the tape passes before reaching the recording head. There is also provision for rapid spooling of the tape in either direction to enable it to be rewound or wound on to cue.

A basic record/replay system. A single motor drives the capstan drive spindle and the storage and take-up spools through a system of pulleys.

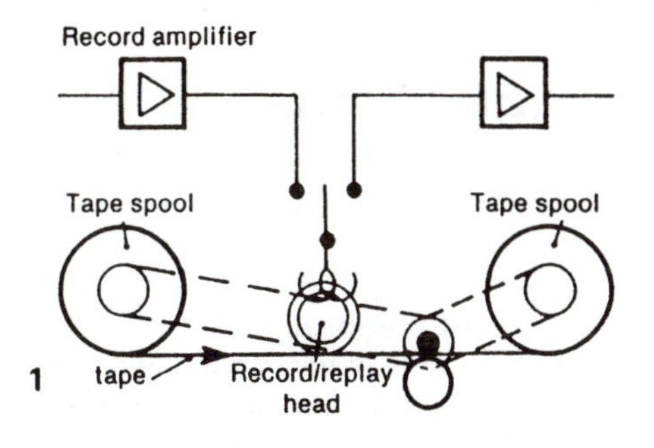

A domestic machine incorporating an erase head (e) but with a shared record/replay head. Tape movement and spooling is controlled by a single motor. The tape is held against the head by a pressure pad.

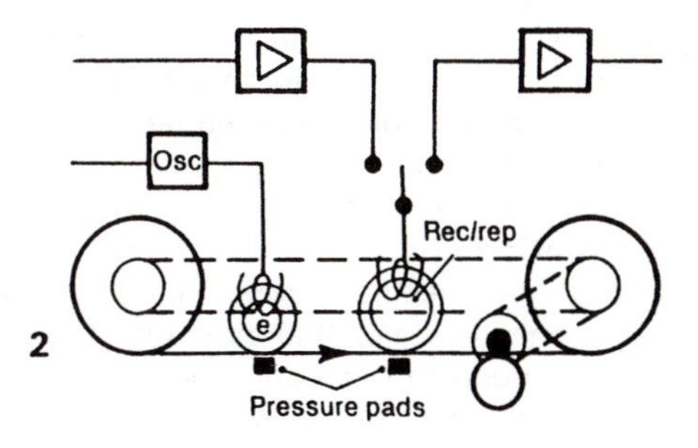

High-quality machines employ separate record and replay heads. Separate motors are used to drive the tape capstan spindle and each of the tape spools. This allows for proper control of tension of the tape against the heads without the aid of pressure pads.

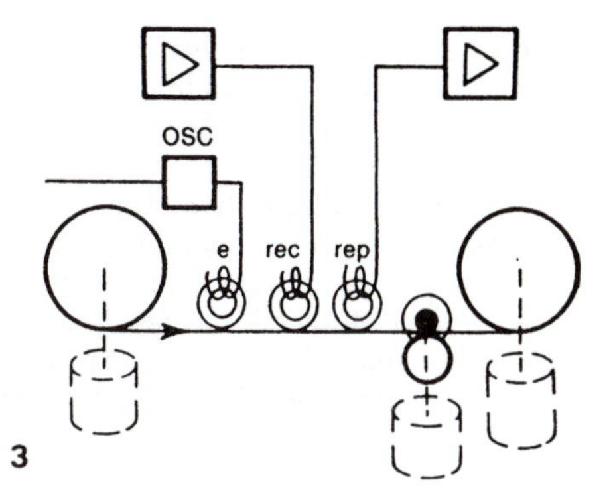

Magnetism is a basic force that occurs as a natural phenomenon in certain minerals.

Magnetism

Magnetic recording is a method of storing information in a material in terms of its state of magnetism.

Permanent magnets

Many hundreds of years ago it was discovered that pieces of natural iron ore called 'lodestones' were found to attract other pieces of iron and to attract or repel other lodestones according to certain laws. Moreover, if they were freely suspended by means of a thread they would turn in a particular direction in relation to the north/south poles of the earth. The material, magnetite, which is an oxide of iron (Fe_2O_4), was particularly pronounced in minerals located in the region of Magnetia, a city in Thessaly – hence the name 'magnet'.

Magnetic poles

Magnetism is found to increase towards the ends of a magnet; the longer and thinner it is the more this is the case. The two ends, called the poles, are nominated north (seeking) and south (seeking) according to which of the earth's magnetic poles they point to when freely suspended.

Magnetic field

If a compass needle is moved in the area of influence of a magnet; (known as its 'magnetic field') and the direction it indicates is plotted, the needle will be found to trace out a series of concentric circles in the directions followed by what are called *lines of magnetic force.* The direction of these lines of force in a magnetic field is taken, by convention, to pass through the magnet in the direction from south to north and outside the magnet from north to south.

Magnetic field strength

The total number of lines of force that form the magnetic field is called the *magnetic flux* (symbol Φ). These lines tend to be unevenly distributed so that the degree of magnetic effect at a particular point depends upon the concentration of flux or *flux density* in that area. Magnetic flux density (symbol B) is proportional to $\Phi \div$ area.

Magnetic effect

If the opposite poles of two magnets are brought together lines of force will link, forming a bond, and they will attract each other.

If like poles are brought together the lines of force, being unable to cross because they represent the resultant of the forces existing at the poles, compress, causing repulsion. The force of attraction or repulsion between two magnets is inversely proportional to the distance between them.

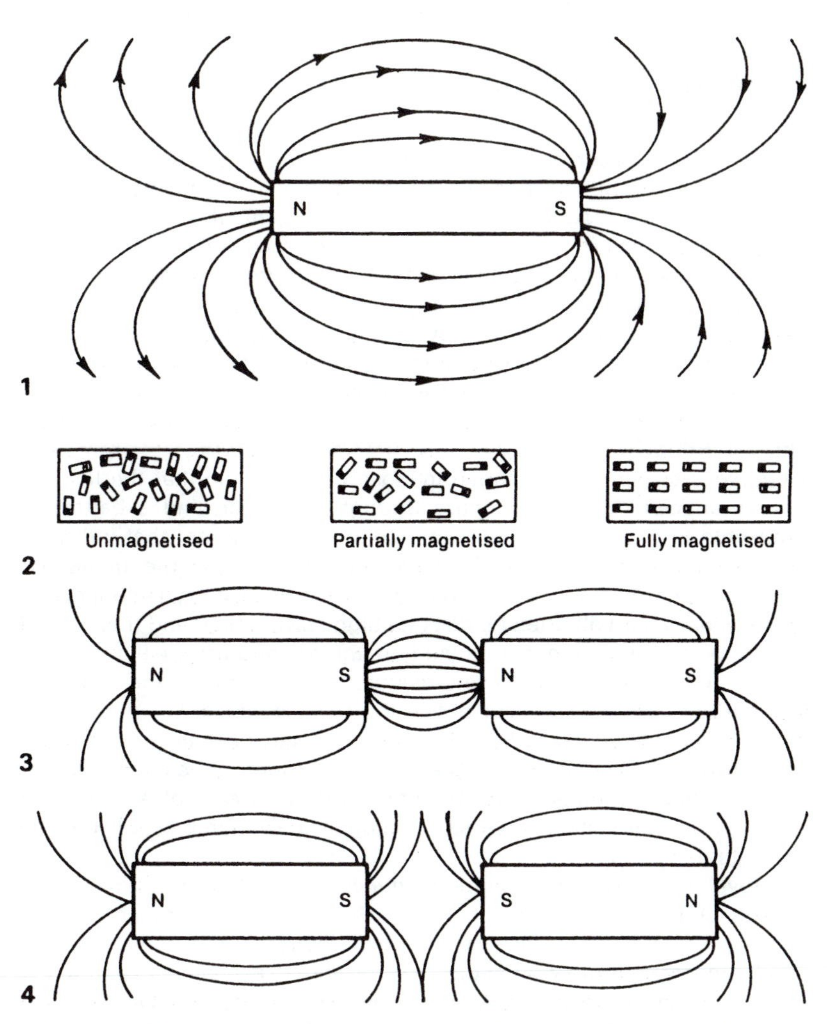

1. Illustrating the lines of force in a bar magnet. The existence of these lines can be demonstrated by placing a card sprinkled with iron filings over the magnet and tapping it lightly. The filings will form up in the manner shown, but of course the effect occurs in three dimensions.

2. Matter is composed of molecules each of which can be considered as a minute magnet. In an unmagnetized substance they are arranged in haphazard fashion and have no overall magnetic effect. In a partially magnetized substance, many of the molecules have their magnetism orientated in the same direction. In a fully magnetized (saturated) state virtually all the molecules are turned the same way and their magnetic effect adds up.

3. Lines of force tend to shorten themselves, i.e. contract along their length. This tends to draw opposite poles together.

4. Lines of force tend to expand laterally, i.e. resist lateral compression; as they cannot cross, like poles repel each other.

A changing magnetic field that cuts a conductor induces into it an emf proportional to the rate of change.

Magnetic Induction

When an electrical current flows in a conductor, lines of magnetic force are set up in and around the conductor. It can be shown that the movement of a conductor in a magnetic field, or the movement of a magnetic field which causes lines of force to cut the conductor, will induce an emf into the conductor and, if there is a circuit, cause a current to flow through it. Thus, a reversal of the process by which information in the form of electrical signals can be stored, by electrically magnetizing an element, can be used to retrieve the information by causing the magnetized element to move in relation to the conductor or coil.

Mutual induction

If two coils of wire are arranged in close proximity, so that the lines of force produced by one also pass through the other, and a battery is connected to one coil and a galvanometer (an instrument that indicates the strength and direction of an electrical current) to the other, the following phenomena can be observed. When the battery is connected, the galvanometer will briefly indicate a high current flow and then return to zero. When the current is switched off, the galvanometer will indicate a brief current flow in the reverse direction. If the polarity of the battery is reversed, the same will occur but in the reverse sense. If the switch is exchanged for a variable resistance a similar reaction can be produced but only while the resistance is being altered; the extent of the galvanometer deflection then depends on the rate of change of the control. Again, if the resistance is altered in the reverse direction the galvanometer will indicate opposite deflection.

The value of the *electromagnetic induction* depends upon the rate of change of the flux linkages between the circuits. The most obvious practical example of the use of this phenomenon is the transformer, in which the two coils are wound on the same core to achieve maximum flux linkage. Instead of varying a direct current, an alternating current is used so that the magnetic field is continually building up and collapsing upon itself with the repeated changes in polarity.

Self induction

If a changing magnetic field around a conductor or coil can induce an emf into an adjacent conductor, it will obviously have the same effect upon itself. Changing the current flow through a conductor produces a change in the magnetic field surrounding it. In cutting the conductor, the change in the field induces an emf which tends to maintain the status quo, i.e. it tends to increase a decreasing current flow or vice versa. This property of induction, which is analogous to inertia in mechanical terms (in that it tends to resist change), is proportional for frequency, i.e. rate of change.

1. In this circuit while the resistance is being adjusted the galvanometer will deflect, showing that the current in the secondary coil is endeavouring to oppose any state of change in the flux linking the two coils.

2. *Time constant.* When a circuit has inductance, the action of switching a current on or off converts electrical energy to magnetic energy or vice versa. This takes a finite time.
In the circuit, if the switch is made to b, were it not for the inductance, when this circuit was switched on the current would rise instantly to the full value determined by the total resistance of the circuit.
Due to the transfer of energy caused by the inductance, the current rises more slowly (the growth curve).
The reciprocal situation occurs if the inductance is short-circuited by the switch. The collapse of the magnetic field induces a current to flow. The time that the current takes to rise to 63.2% of its final steady value is the *time constant* (which equals inductance ÷ resistance).

3. The effect of passing current through the conductor is to cause an increase in the flux on one side and a decrease on the other, giving rise to a mechanical force (F) in the direction indicated.

4. Induction through movement in a magnetic field. If the conductor is moved at right angles to a magnetic field, a current will be induced in the direction that produces a force which tends to oppose the motion (see above).

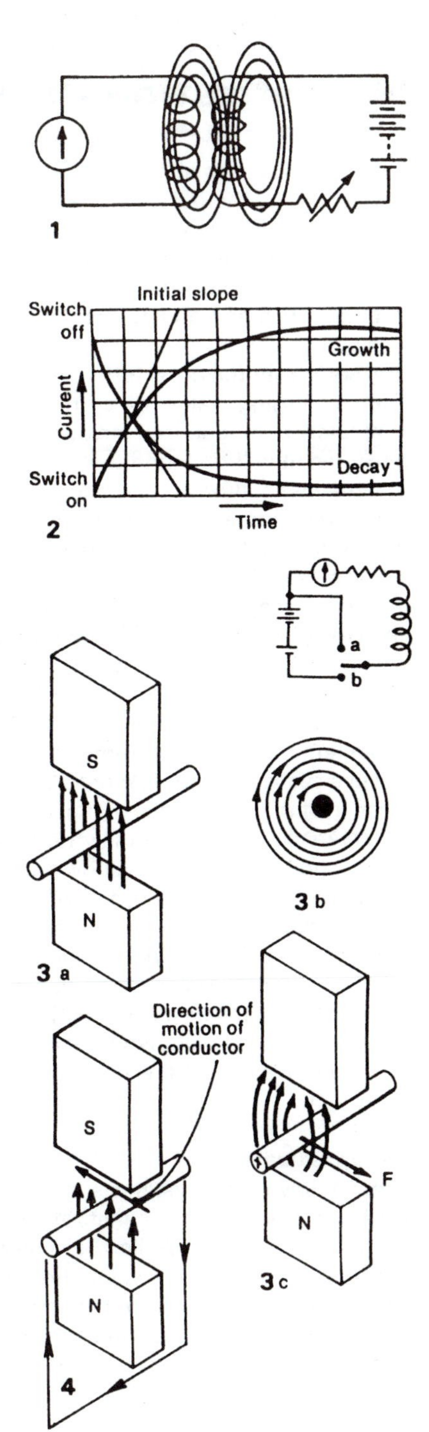

When a ferrous material is magnetized alternately in opposite polarity the magnetizing characteristic does not retrace itself but forms into a 'hysteresis' loop.

Magnetization Characteristic

On page 142 it was shown that a piece of ferromagnetic material can be magnetized by the passage of an electrical current through a coil of wire surrounding it. The extent to which the material remains magnetized after the current is switched off depends upon the retentivity of the material. This is high for hard magnetic materials such as steel and low for soft materials such as iron and permalloy.

Magnetizing characteristic

If an increasing value of current is applied to a coil surrounding a completely unmagnetized ferrous material and the induced magnetism is measured, it will be found that the relationship is not a straight-line one. The magnetic flux at first increases only slowly with increments of magnetizing force; it then increases over the main part of the curve, before flattening off to the *saturation* value when the material is fully magnetized. With some substances, notably ferrous oxide (which is an important material in magnetic recording), the flux density does not return to zero when the magnetizing force is removed at saturation but changes to some higher value which represents the *remanence* of the material.

Hysteresis

If an alternating current is applied to a coil so that the magnetizing force acts in a cyclic manner, magnetism will be induced in alternating polarity. Because of the remanence of the material the magnetizing/demagnetizing characteristic does not retrace the initial magnetizing curve since a force in the opposite direction (known as the *coercive* force) is required to reduce the flux density to zero. After a number of repeated reversals the curve settles down into a loop formation known as a *'hysteresis loop'* (from the Greek *husteros* – coming after). This characteristic, which involves magnetizing the material to saturation in alternate polarity, is known as a *major hysteresis loop*. Any small repetitive reversals in the magnetizing force will give rise to *minor hysteresis loops*. These appear like small versions of the hysteresis curve attached to the inside of the major loop.

Demagnetization

From a study of the hysteresis loop it would appear that, once saturation had taken place, zero magnetization could never be attained with zero magnetizing force. If while the alternating force is being applied it is progressively reduced in value, however, a series of diminishing hysteresis loops are produced which eventually reach zero. This process is called *demagnetization* or, in some circumstances, *erasure*.

Variations of flux density (B) with magnetizing force (H). In fine detail the curve is not smooth but is made up of a series of minute steps (called Barkhausen jumps) as the magnetic crystal elements, which form 'domains' that act like tiny bar magnets, jump into line.

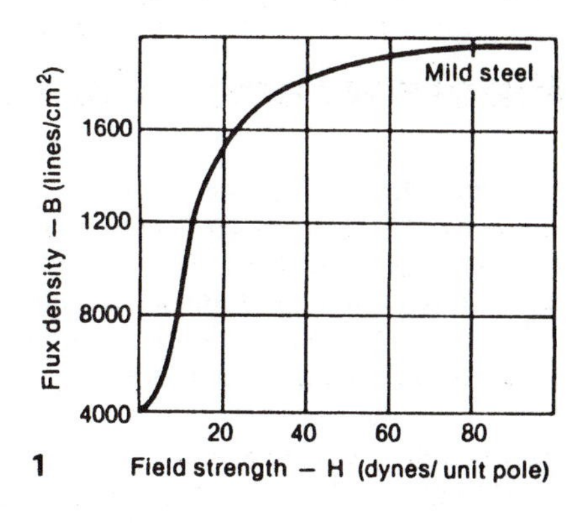

1

Curve a shows how the flux density in a medium increases from zero with increasing magnetic force. Curve b shows the effect of progressively reducing the force to zero. The flux density will diminish not to zero but to a value (r) which is the *remanence*. If the sense of the magnetizing force is reversed (c), a point will be reached where the force exactly balances the residual magnetism of the material. The value of this force ($+H_c$) needed to achieve this state from saturation is the *coercivity* of the medium.
The material is not completely demagnetized at this point and if the force were removed the flux density would rise as in curve d. A further increase in +H is required (to $+H_d$ curve e) to completely demagnetize the material.

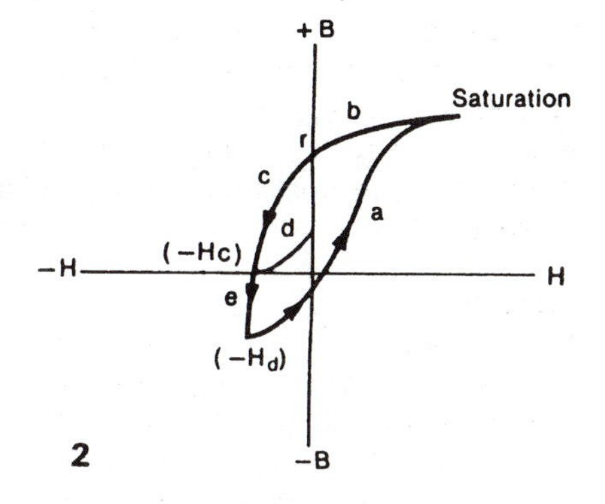

2

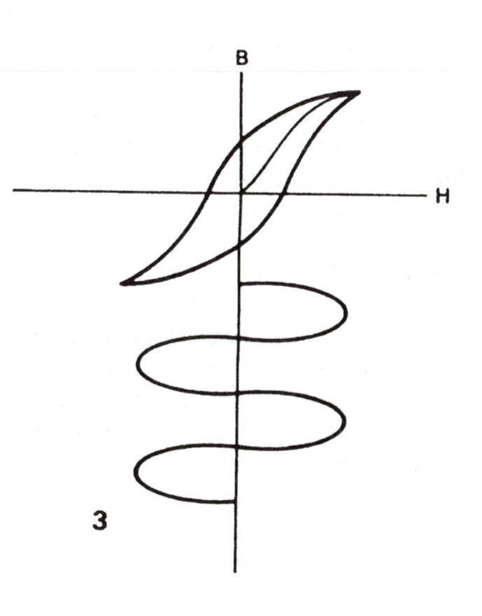

3

The initial magnetization and subsequent variations of flux density caused by sinusoidally varying the magnetizing force following many reversals.

Magnetic fields can be produced by passing current through an electrical conductor.

Electromagnetism

Electromagnetic fields

When an electrical current flows in a conductor the motion of the electrons sets up a magnetic field around it. This takes the form of concentric lines of force within and outside the conductor.

The direction of these lines in relation to the current is always such that they can be considered as rotating clockwise when the current is flowing away from the observer.

Magnetic coils

If a conductor is wound into a coil the lines of force will amalgamate to produce a continuous field similar to that of a bar magnet (while the current remains on) with a north pole at the end where the lines of force emerge and a south pole where they re-enter the coil. This forms what is called an electromagnet or solenoid.

Magnetizing force

The strength and disposition of the magnetic field in a coil depends on the number of turns, the current flowing in them and the size and shape of the windings.

$$\text{Field strength, H} = \frac{In}{\iota}$$

Where n = the number of turns, ι = length of coil and I = current. If the coil is formed into a continuous loop or toroid there will be no poles, most of the flux will be contained within the windings and the density will be practically uniform throughout the cross section.

Permeability

The efficiency of a coil in terms of the magnetic effect, i.e. number of lines of force produced by a given magnetizing force, can be considerably increased by winding it on a core of ferrous material. This efficiency (permeability) can be expressed as the ratio of the magnetic induction or flux density (B) to the field strength (H).

$$\text{Permeability, } \mu = \frac{B}{H}$$

In the gauss system the permeability of free space (vacuum) is taken as unity. This is known as absolute permeability, μ_o. The ratio of the permeability of other materials to absolute permeability is called the relative permeability, μ_r, and varies considerably between materials. It is very small, and substantially constant, for air and for such *paramagnetic* materials as copper, aluminium etc. *Ferromagnetic* materials such as iron, steel, cobalt, etc. have very much higher and more variable values of μ (orders of magnitude of 200–200 000) depending upon their composition, purity and temperature.

When current flows in a conductor, a magnetic field is set up around it. The lines of force flow in a clockwise direction as seen from the source from which the current flows.

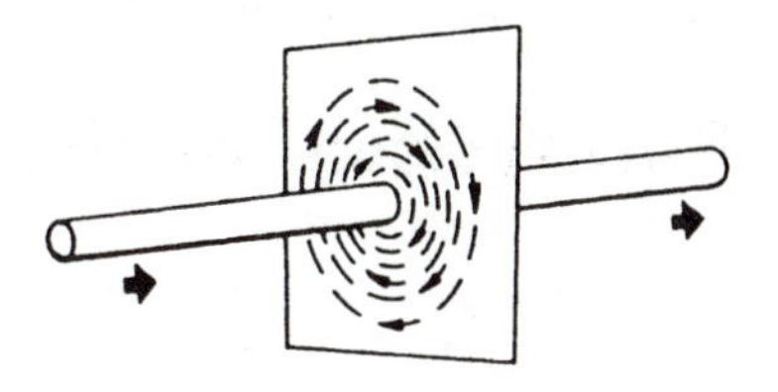

If a conductor is wound into a coil, the lines of force will amalgamate to form a continuous field similar to a bar magnet.

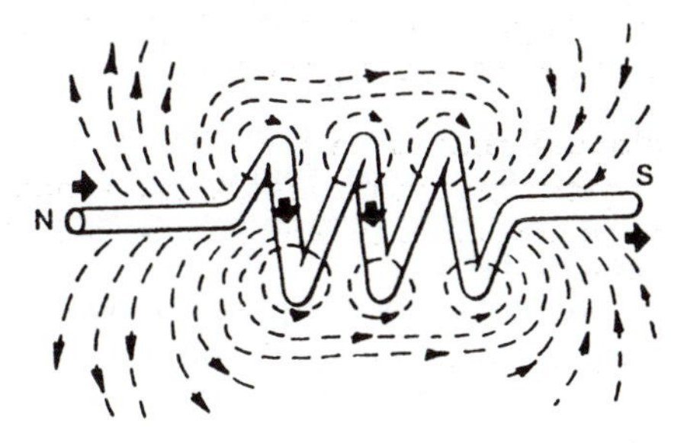

The efficiency of a coil is increased by winding it on a core of ferrous material. The extent to which the core material improves the magnetic efficiency depends upon its *permeability*. This is the ratio of flux density to field strength. The graph shows a typical permeability curve for mild steel.

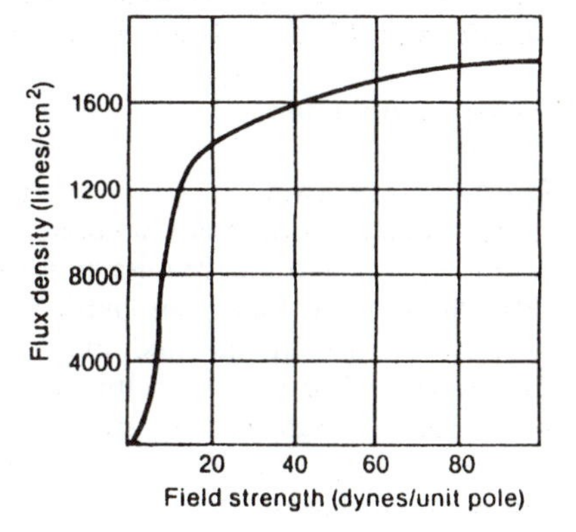

Recording tape is manufactured in a variety of sizes and thicknesses.

Recording Tape

The basic material for magnetic recording is tape. Magnetic disc can also be used but this tends to be reserved for specialist applications such as digital systems and computer memories.

Tape sizes

Recording tape is manufactured in wide rolls and slit into various widths to suit different purposes. These include:

0.15 in (3.81 mm) for compact cassettes
0.25 in (6.2 mm) the standard for most reel-to-reel machines
0.5 in (1.27 mm) for multi-track audio and television recording
1.0 in (2.54 cm) for multi-track audio and television recording
2.0 in (5.08 cm) for multi-track audio and television recording

There are also 35 mm and 16 mm recording tapes with sprocket holes to match film for the synchronous recording of film sound, and the magnetic stripe, which is a thin coating of magnetic material applied to one edge of a cine film on which sound can be recorded.

Base material

Recording tape is composed of two layers: the base and the coating of oxide on which the recording is made.

Various materials are used for the base. Early examples were made of cellulose acetate, which was inclined to absorb moisture and dry out with storage, becoming brittle.

This has been superseded by PVC and later polyester. A tensilized polyester, which has improved tensile strength, is particularly suitable for the thin double- and triple-play tapes. The requirements for base materials are freedom from:

Stretch Stretching of the tape can give rise to variations of pitch in the recording.

Cupping A tendency to bend across its width can result in poor head contact unless pressure-pads are used.

Bias One side of the tape being longer than the other causes poor spooling.

Curl A tendency for the edges to ruffle can cause pitch variations.

Spooling systems

Reel-to-reel The tape is stored on a spool and is played by winding it from the storage spool to a take-up spool.

Cassette The tape runs between two spools, to which it is permanently attached, fitted into a plastic box.

Cartridge The tape is in the form of a continuous loop wound on a single spool with rollers in a single box.

Domestic reel-to-reel plastic spool.

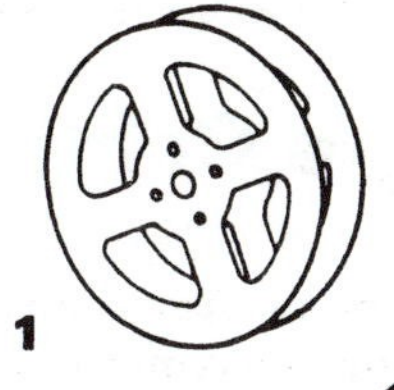

1

Professional recording is done on large spools. In the cases where ¼ in tape is used the spools are usually 10½ in (27 cm) NAB spools, usually made of aluminium and sometimes with single face-plates for horizontal mounting.

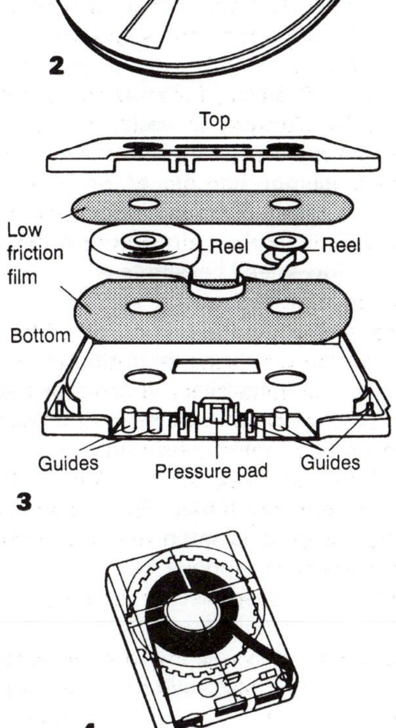

Compact cassettes use tape only 0.15 in (3.81 mm) wide permanently attached to a supply reel on the left and a take-up reel on the right. The tape is carried on spools with no retaining cheeks. The tape is guided past the head by plastic pins set into the cassette case, and a pressure pad on a steel strip spring holds the tape against the head.

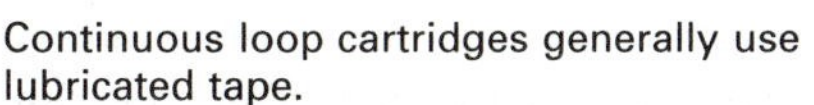

Continuous loop cartridges generally use lubricated tape.

The R-DAT cassette is totally enclosed because the tape is only 13 μm thick and must not be handled.

Loading grip
Lower shell
Upper shell
Slider
Lid
Notches for slider and incorrect insertion prevention
Notches for slider lock release

5

There are various types of tape coatings, each having different characteristics.

Recording Tape Coatings

The search for improved frequency response and better signal-to-noise ratio has resulted in the production of various types of tape coatings, each with different characteristics.

Ferric oxide (Fe_2O_3) coating

The iron oxide particles (Fe_2O_3) are pulverized into needle form (about 0.5 μm long) and added, in suspension, to an adhesive compound.

Ferric oxide tape tends to have a limited high frequency response and rather poor signal-to-noise ratio.

Chrome dioxide (CrO_2) coating

Chrome dioxide particles have a finer needle shape than ferric oxide allowing for better homogeneity. Chrome dioxide is distinguished by its dark colour (almost black). It provides an excellent high frequency response with the ability to record very short wavelengths (approximately 10 dB better than ferric oxide at 10 kHz). This makes it particularly suitable for slow-speed cassettes.

Chrome dioxide has excellent power-handling capacity in the high frequencies but a fairly high inherent hiss level. To get the best out of it, therefore, the input high frequency and bias level should be high and the replay characteristic almost a pure 6 dB/octave (see p. 166). Unfortunately chrome dioxide tape tends to produce a high degree of head wear and has a slightly inferior low frequency response compared with Fe_2O_3.

Ferrichrome (CrO_2. Fe_2O_3) coating

Ferrichrome is intended to combine the virtues of the enhanced high frequency response of chrome dioxide with the low and medium frequency response of ferric oxide. A thin layer (1 μm) of chrome dioxide is superimposed over a ferric oxide base, approximately 5 μm thick. Unfortunately, as the chrome layer is on the outside there is still the problem of excessive head wear due to abrasion.

Cobalt coatings

Substantial improvements have been made in frequency response, signal-to-noise ratio and dynamic range without the need for critical bias settings, by the addition of cobalt to ferric oxide. This can be achieved by introducing an ion of cobalt into a particular formation of ferric oxide in a process similar to the 'doping' method used in the manufacture of semiconductor materials. The surface retains the relatively smooth, soft texture of ferric oxide so that head wear is not excessive.

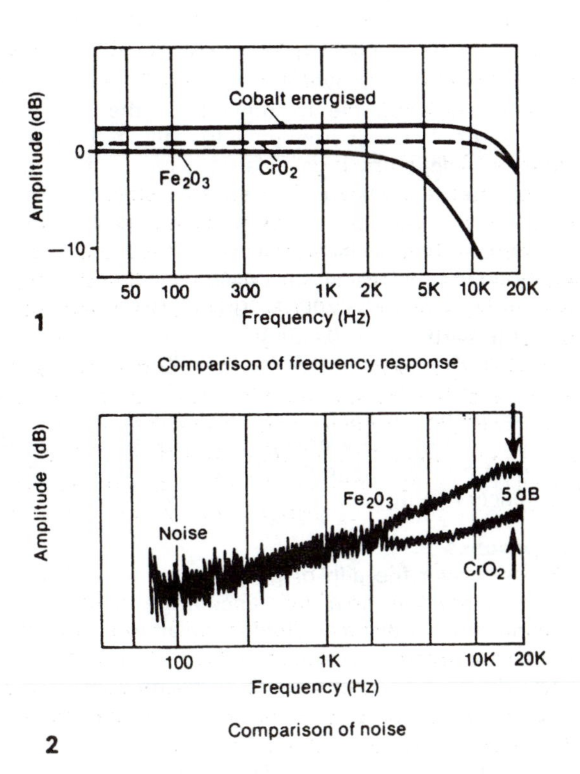

1. The graph shows a comparison of frequency response characteristics between ferric oxide, chrome dioxide and cobalt energized.
2. Comparison of noise spectrum between ferric oxide and chrome dioxide tapes.

Previously recorded material is erased from the tape by applying a powerful high-frequency alternating magnetic field.

Tape Erasure

Before a recording is made, it is usual, unless a superimposition is required, to subject the tape to a full erasure to ensure that no unwanted modulation remains on it.

For this reason a special erase head is fitted on most machines and is the first head that the tape encounters.

The erase head

The erase head consists of a coil of wire wound on a ring of magnetic material which has a small gap over which the tape passes. When erasing is required (normally when the machine is in the record mode), an alternating current of ultrasonic frequency (usually in the region of about 50 kHz to 100 kHz), supplied by an oscillator, is passed through the coil.

Quite a high power is involved as it is necessary to bring the tape to magnetic saturation, so the core is usually made of some very low coercivity material such as silicon steel which does not saturate readily. The core is built up in thin laminations, each insulated from the others or made of ferrite, a nonconducting magnetic material, to reduce the formation of eddy currents in the metal and consequent losses due to heat. The head gap is filled with a nonmagnetic substance such as hard copper or nickel. Eddy currents are encouraged to occur in this area as they tend to assist in the erasing process. To combat the high coercivity of modern recording tapes, many erase heads are produced with double gaps consisting of a small magnetic element sandwiched between two nonmagnetic spacers so that the tape is effectively wiped twice.

The erasing process

As the tape passes over the gap or gaps in contact with the poles, the magnetic flux circulating in the core is encouraged to divert through the tape due to its relatively low permeability. The alternating magnetic force is sufficiently strong to drive the tape around its major hysteresis loop (see p. 155) from saturation in one direction to the other.

Erase heads have quite wide gaps relative to the wavelength of the bias (typically 0.1–0.5 mm) so it is possible for an element of tape to be subjected to a number of alternations of magnetism as it traverses the gap. As each element of the tape passes the head it is subjected to an alternating magnetizing force which builds up to saturation and then diminishes to zero. The tape magnetization therefore follows a series of increasing and diminishing hysteresis loops. The magnetic properties of the molecules are continually reorientated in opposite directions until, at the end of the gap, the flux is no longer strong enough to affect them. They are, therefore, left in random formation and the tape is demagnetised.

Double DC erasure

Some digital tape recorders employ two DC erase heads in tandem. One takes the tape to saturation in one direction, effectively wiping any pre-recorded material, and the other works in the opposite polarity, bringing the tape back to a nearly demagnetized state.

The erase head. The double head gap consists of a central magnetic section sandwiched between two nonmagnetic spacers. This has the effect of virtually erasing the tape twice thus ensuring complete erasure. An ultrasonic alternating current (supplied by an oscillator) is passed through the head coils. The current is powerful enough to take the tape to magnetic saturation in each direction. The head core is composed of finely laminated soft iron or ferrite to reduce losses due to eddy currents.

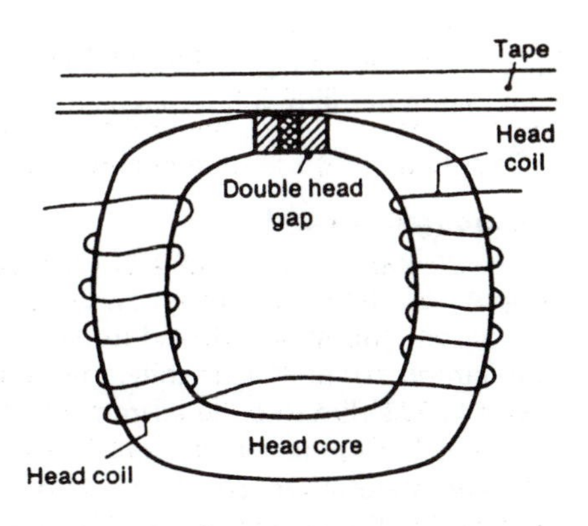

The erasing process. The tape passes through an AC magnetic field that takes it to saturation in alternate polarities. As each particle of tape leaves the gap it passes through a diminishing field so that the hysteresis loop, which has built up to saturation value, collapses upon itself and the residual magnetism is reduced to zero. The tape is thus demagnetized. Point a shows the expanding flux, b the flux linkage diminishing and c the expanding and contracting hysteresis loop, reaching saturation at the mid-point of the gap so as to absorb any residual magnetism from a previously recorded audio frequency signal.

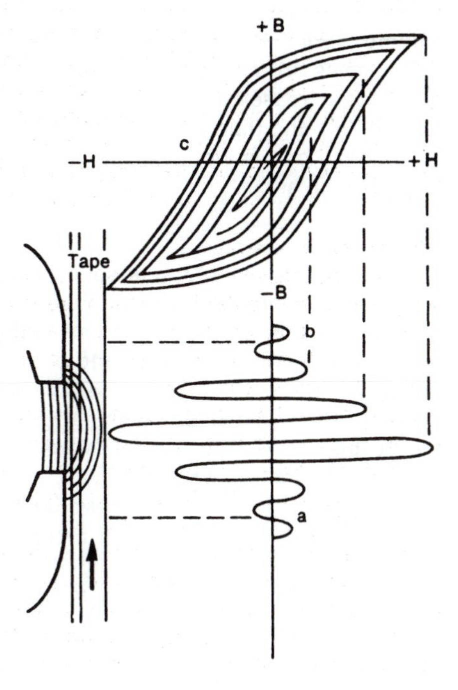

The way in which electrical signals are transferred to magnetic variations on tape varies according to frequency.

The Recording Process

The purpose of recording is to impose on the tape a permanent magnetism that varies in polarity and intensity along the tape in direct relationship to the manner in which the signal being recorded varies with time.

The recording head

The recording head is a ring of ferrous material composed of ferrite, a compound with a high resistivity, or of a material with a high permeability such as *permalloy* (an alloyed iron containing silicon). The material is in granular form or built up in thin laminations, each individually insulated to reduce losses due to eddy currents.

The ring is made in two halves joined together with shims of non-magnetic material which fill the gaps at the front and back. Some high quality recording heads have quite substantial back gaps which, although they reduce the overall permeability and therefore require more driving power, make the recording more consistent throughout the frequency and volume range.

The poles are chamfered at the back of the gap over which the tape passes to increase the reluctance of the gap and help to push the magnetic flux out into the tape.

The head coil is wound in two sections (one on each leg) which are connected in series.

Gap size

Various factors, notably tape speed, affect the choice of optimum head gap size for recording. It needs to be sufficiently large to produce a strong enough external field and deep enough penetration of the tape. On the other hand, if it is large enough to be comparable with the wavelength of the signal, loss of the high frequency response will result. This is because the magnetizing force, diminishing towards the end of the gap, will cause the AC bias (p. 158) to act in the manner of the erase signal and cause partial demagnetization.

A typical recording head gap would be of the order of 0.02 mm. Where only one head is used for both recording and reproducing, as in many domestic machines, the gap is made to suit the more stringent requirements of replay. This requires a much narrower gap of the order of 0.002 mm.

The recording process

The manner in which the recording is established on the tape is very complicated and varies with frequency. Different segments of the transfer characteristic become involved as the frequency and thus the relationship between wavelength and head gap size varies.

The recording head is made of thin laminations of ferrite or iron alloy, each insulated from the other to reduce eddy current losses. The back of the front gap is chamfered to increase the reluctance. It also helps to push the magnetic flux out so that it penetrates the tape more deeply.

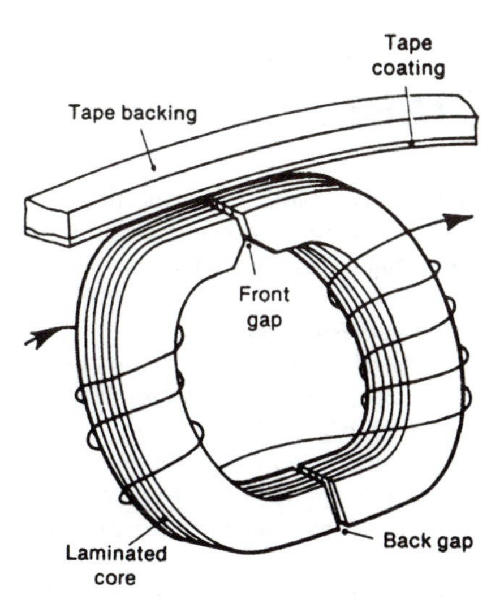

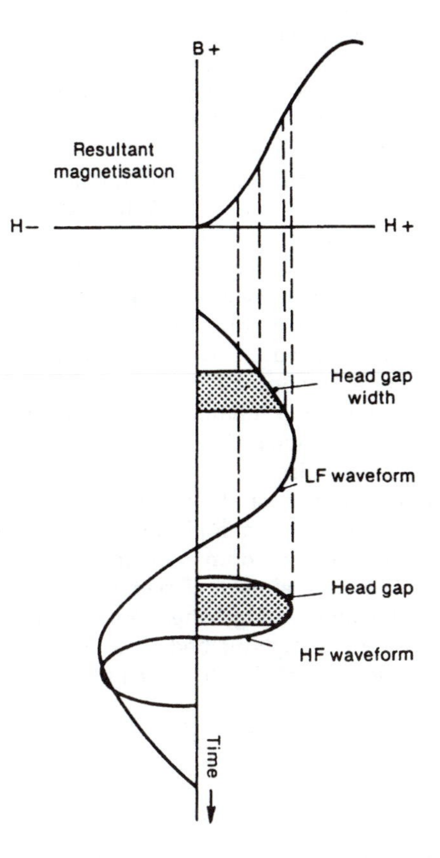

The magnetization transfer characteristic.

At low frequencies the gap represents only a small segment of the waveform so that the magnetizing force is virtually unidirectional and the output tends to relate to the initial transfer characteristic. As this is non-linear, harmonic distortion results. At higher frequencies, where the wavelength is less than about eight times the gap size, reversals of direction of magnetization will occur while an element of tape is passing the gap. This results in the formation of incomplete hysteresis loops with consequent distortion.

DC Bias

Having considered the properties of the tape and the characteristics of the recording head, we should now consider the way in which the programme signal becomes translated into a magnetic recording on the tape.

The need for bias

A study of the curve on page 145 shows that the transfer of alternating electrical energy into magnetic induction is by no means a linear relationship.

The initial magnetization characteristic is in the form of an S in either direction from zero. If, therefore, the signal was merely applied to the recording head, severe distortion of the recorded waveform would result.

The nature and extent of the distortion would vary considerably with frequency and amplitude of the signal in relation to the tape speed and head gap size, as explained on the previous page.

Use of DC bias

The problem of non-linearity of the transfer characteristic is tackled in some simple recording machines by the application of a magnetizing force of fixed value and polarity which can be used in either of two ways:

1. A direct current is superimposed on the signal to bring the operating point to the centre of the straight line portion of the transfer characteristic.
2. The tape is brought up to full saturation by a previously sited erase head (possibly even a permanent magnet) to ensure complete erasure of any previous magnetization. The signal is then superimposed on a small demagnetizing DC current so that it follows the straight portion of the hysteresis loop.

This second method has the advantage over the first one that it pre-erases the tape but both systems only work effectively for low frequencies and the noise level is very high due to the varying effect of the DC magnetization on the random particles in the tape.

As the tape coating is composed of a variety of different sized particles orientated in random manner, it is not difficult to imagine that the DC field will produce a residual magnetism with variations matching the structure of the tape coating. When reproduced these variations will be heard in the form of a high-pitched hiss. For this reason it is usual for systems employing DC bias to curtail the high frequency response.

By superimposing a small fixed DC current on the audio frequency signal, the signal can be shifted to the straight part of the transfer characteristic thus reducing distortion.

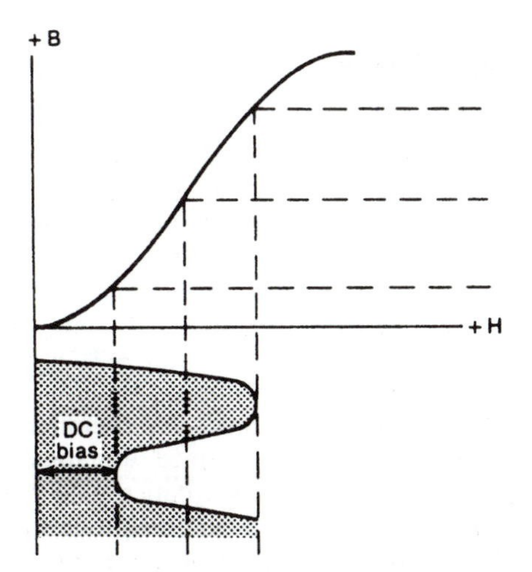

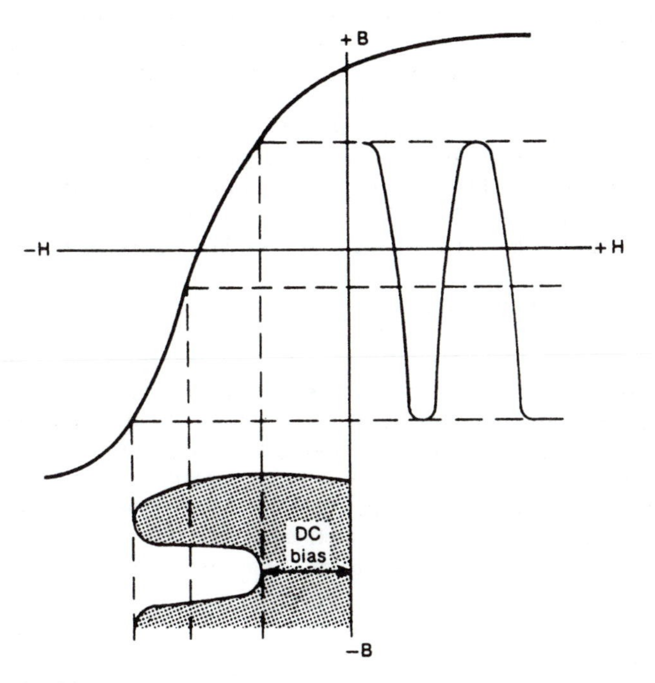

Saturation bias.

A strong bias is used to bring the tape up to saturation. This erases any previous signal recorded on the tape and sets it to its major hysteresis curve where there is room for a larger signal excursion. A larger signal can thus be recorded and the programme:noise ratio is improved.

The use of AC Bias reduces distortion and allows for high output.

AC Bias

We have seen (p. 145) that, due to the non-linearity of the transfer characteristic, it is necessary to apply bias to the signal when recording. The use of DC bias results in a high level of background noise. It has been found that by using an ultrasonic AC (typically about 75–100 kHz) as bias much better results can be obtained and this is used in practically all modern tape recorders.

The effect of head gap

Firstly it is important to remember that the intention of magnetic recording is to induce on to the tape a residual magnetism which varies along its length in accordance with the signal to be recorded. In other words what matters is the magnetic state of each particle of the tape as it *leaves* the influence of the head gap.

In discussing tape erasure we found that applying an AC field to the tape, which increases and then diminishes in effect as it crosses the gap, produced a series of hysteresis loops which expanded to saturation and then collapsed progressively, thereby reducing the residual magnetism to zero.

If an AC field is applied which is not strong enough to take the tape to saturation in each direction and on this AC field we superimpose a DC field, this will have the effect of shifting the general location of the hysteresis loop. As far as each individual particle of tape is concerned, as it leaves the gap the AC field of the bias will collapse not to zero but to the DC value which forms the centre of the shifted hysteresis loop. Hence the tape is left with an appropriate degree and direction of residual magnetism.

The effect of wavelength

Note that this example differs from the case of superimposing an AC programme signal on a DC bias because the wavelength of the AC bias is much shorter than the head gap size so it has time to form complete hysteresis loops as each particle traverses the gap. If, instead of the DC example, we superimpose an audio frequency signal on the AC bias, it behaves like a slowly varying DC as far as each particle of tape crossing the gap is concerned because its wavelength is long in relation to the head gap (at least ten times longer than the bias). A residual magnetism is left that conforms with the audio waveform.

Anhysteretic curve

By superimposing the programme signal on an AC bias we have effectively absorbed the non-linear element of the magnetization/demagnetization process in the bias. The transfer characteristic now follows what is known as an anhysteretic curve.

The effect of superimposing a DC magnetism on an AC magnetizing field. If the DC is suddenly applied, it takes a few alternations before the hysteresis loop settles down into its new position (displaced to the value of the DC bias).

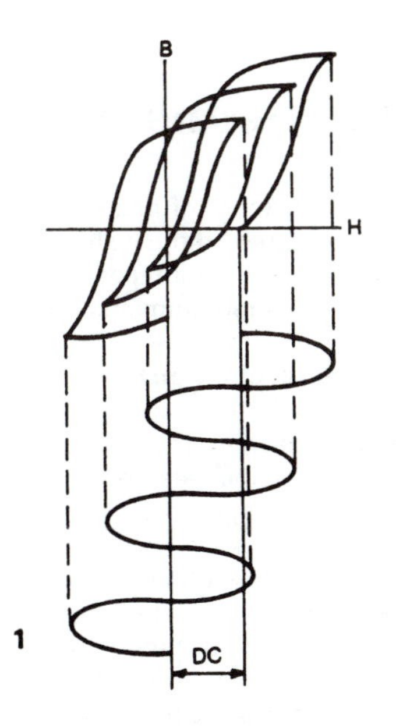

A representation of the magnetizing effect on the tape as it passes the recording head gap. a, If only the AC bias is present, it virtually erases itself as it leaves the gap. b, If a DC force is superimposed, the tape is left with a residual magnetism. The audio signal can be considered as a DC source in this context because, having a much lower frequency (longer wavelength) than the bias, it does not have time to execute a complete hysteresis loop while traversing the gap and leaves it with a residual magnetism which constitutes the recording.

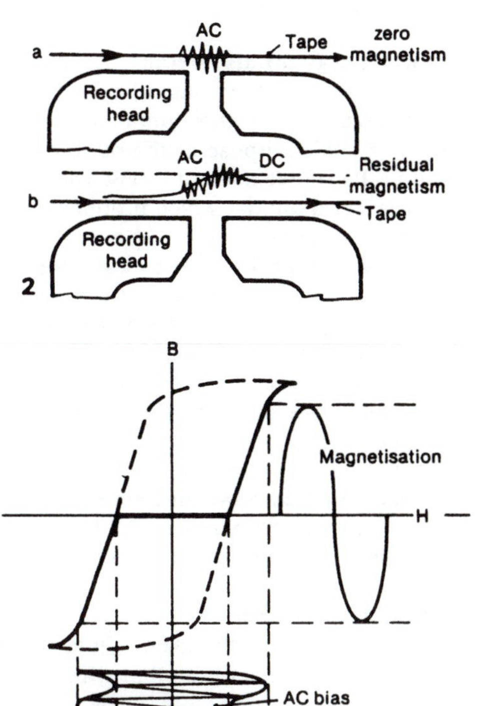

When a ferrous substance is subjected to a slowly changing DC force superimposed on an AC magnetizing force, the transfer characteristic follows an hysteresis curve. The steep slope of this curve helps to explain why the use of AC bias greatly increases the reproduced output.

Optimum adjustment of bias is an important factor in producing high quality recordings.

Bias Adjustment

The correct adjustment of bias level is an important prerequisite of the making of a good recording. It should be checked at regular intervals and re-adjusted when different types of recording tape are used if, as is often the case, their bias requirements differ.

Effect of bias level on output

The strength of the bias current controls the recording sensitivity and therefore has a bearing upon the eventual replay output.

If the relationship between bias current and replay output is plotted for a constant level input, it will be found that the sensitivity increases rapidly with current to a peak value, at which it levels off and then progressively decreases. This is due to the fact that the field laid down by the recording head and the sensitivity of the replay head is greatest at the surface of the tape and falls off with distance, i.e. through the depth of the tape.

Overbiasing the recording head tends to push the region of maximum signal intensity deeper into the tape so that the flux available to the replay head diminishes.

Effect of bias on distortion

To assess the effect of bias on distortion it is necessary to consider this factor in relation to the output level. Maximum undistorted output is defined for this purpose as the output produced by a pure tone in which the third harmonic content is 3%. This condition occurs when the bias is rather larger than that which produces maximum output.

Effect of bias on frequency response

Increasing bias current beyond the value corresponding to optimum output has a progressively adverse effect on the high frequency response. This can be due to the effect, mentioned earlier, of the area of maximum magnetization being driven deeper into the tape. The high frequency response is particularly susceptible to separation between tape and head gap (see p. 158).

Effect of bias current on noise

Tape noise tends to increase with bias current to a point similar to that of maximum sensitivity and then decline sharply.

Choosing bias current

Taking account of all the factors, it is evident that the optimum value for bias is slightly greater than that which produces maximum output.

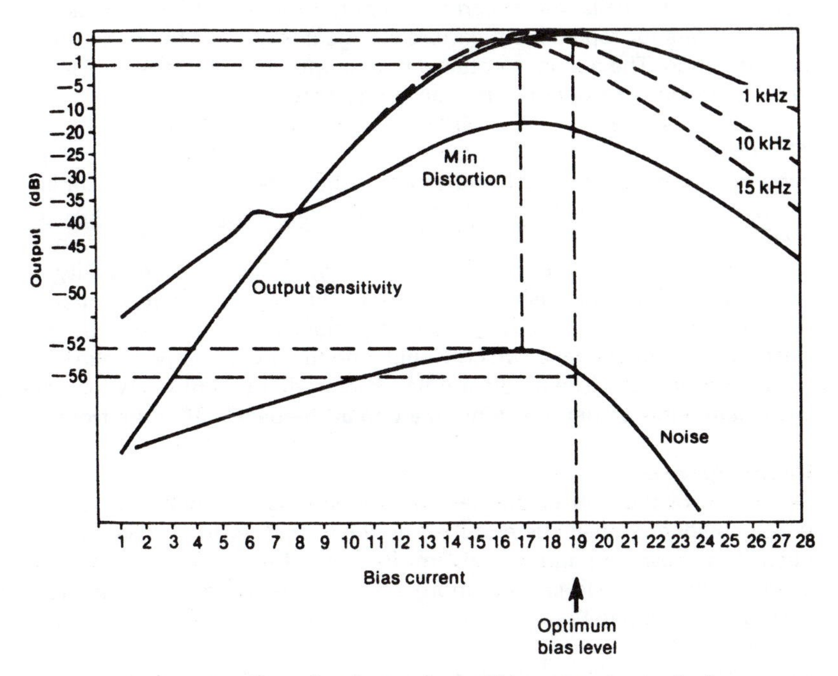

Curves indicating the effect of various values of bias current on output sensitivity, distortion (maximum undistorted output) and noise.

The values for bias amplitude are arbitrary units; the actual level depends on the properties of the tape, the construction of the head etc.

Superimposing the curves makes it evident that the optimum condition for bias occurs at a level just above that which provides maximum output. The method for adjusting bias, therefore, is to set it for maximum output, using standard level tone (1 kHz), and then increase it until the level drops by 1–2 dB. The higher figure gives a measure of insurance against wide fluctuations in level caused by variations in the tape coating. A better system is to use 10 kHz tone, where this is available, and adjust for a 3 dB drop from maximum output with increasing bias. This is because the steeper fall of the 10 kHz curve makes it easier to judge the correct adjustment.

The Replay Head

The replay head is very similar in construction to the recording head; in fact many domestic tape recorders employ the same head for both purposes. This calls for a certain compromise in design but as the requirements for replay are the more stringent this tends to be the prime consideration. The aim is to obtain a high output from the tape consistent with a sufficiently extended frequency response.

The core
Ring-type cores are used. They are composed either of high permeability material made up in thin laminations (of the order of 0.12 mm or less) each insulated from each other or of a ferrite material (a compound containing particles of iron compressed into shape) which has high resistivity. In either case the purpose is to reduce losses due to the formation of eddy currents. These are electrical currents, induced in the core by the fluctuating magnetic field, which would tend to waste energy through the production of heat. The cores are manufactured in two halves which are then clamped together, resulting in a gap at the back and at the front.

Back gap
The effect of the gap at the back of the head is to reduce the overall permeability characteristic and thereby the output. It does make it more consistent, however, and thereby improves the frequency response. The output from the head can be improved by interleaving the laminations across the back gap.

The coil
The coil is usually wound in two sections, one on each leg of the core. They are wound astatically, i.e. connected in series in such a manner that the current caused by induction within the core adds up, whereas that from external fields cancels out. The whole is enclosed in a mumetal shield to reduce interference from external magnetic fields to a minimum.

The front gap
The width of the front gap is related to tape speed in determining the frequency response. As explained on p.154, the high frequency response is limited by the width of the head gap. If a good high frequency response is to be obtained with a relatively low tape speed the gap must be very small. Gap sizes of 0.005 mm or less are common, reducing to 1 μm for some cassette recorders with tape speeds of 4.75 cm/s. The lower limit of gap size is conditioned, apart from practical considerations, by the need to maintain sufficient output at the low frequencies where the proportion of the waveform, and thus the available output, across the gap is reduced.

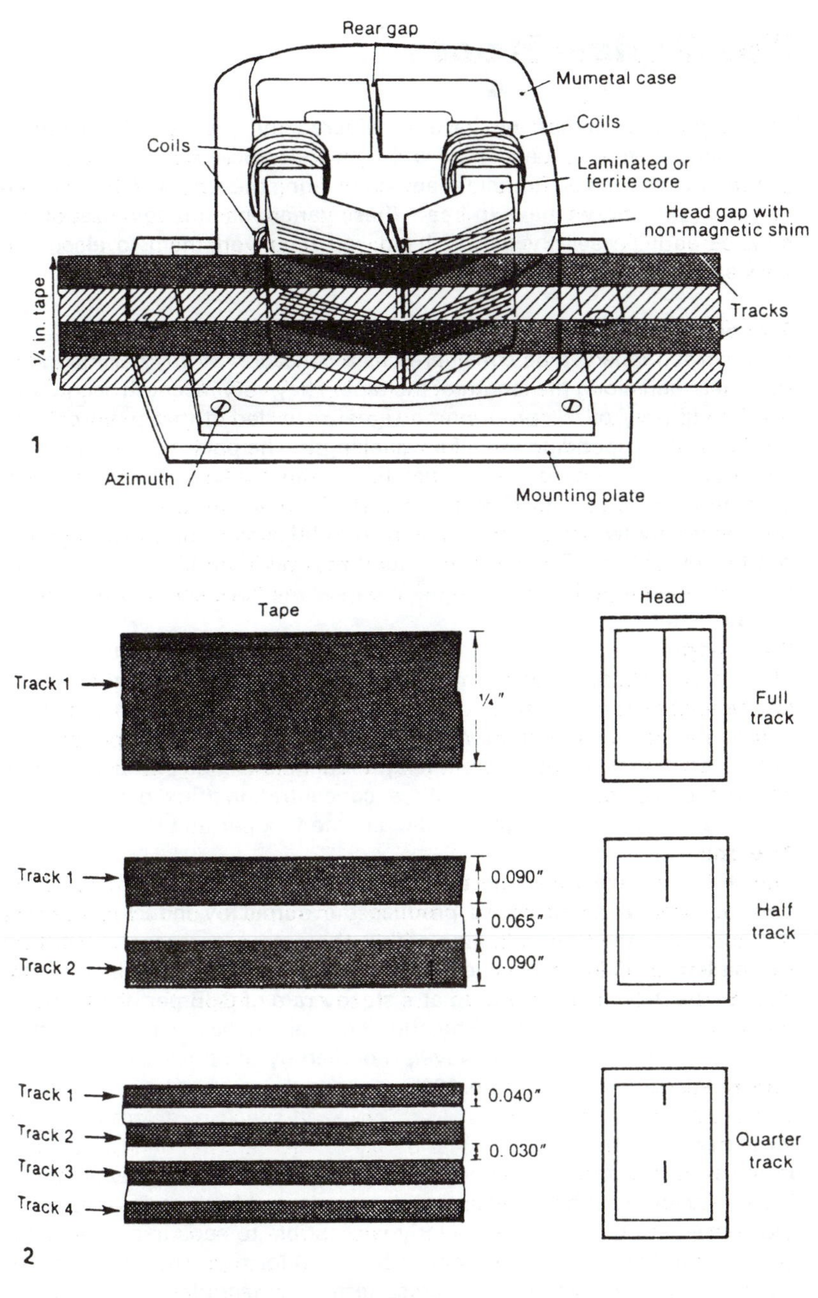

1. A cutaway view of a quarter-track head.
2. Dimensions and positions of tracks on ¼ in tape.

Replaying a magnetic recording is the inverse of the recording process.

The Replay Process

In the preceding pages we have considered the process of magnetic recording, i.e. the conversion of a varying electrical signal into corresponding variations of remnant magnetism along the length of the tape. In the replay process we have to 'read' these variations of magnetization of the tape as it passes the replay head and reconvert them to electrical signals.

The recorded signal

The recorded tape can be considered as a collection of tiny bar magnets laid end to end along the length of the tape. They vary in length according to the frequency of the programme signal recorded at that point on the tape, each corresponding to half a wavelength. The polarity of the field in each magnet is related to the direction of the current that originally induced it, so that a complete cycle of the original waveform would be represented by two magnets with like poles adjacent (north to north and south to south according to the phase of the waveform).

The external magnetic field

As the magnets have open poles the magnetic flux extends outside the magnets and, in fact, outside the surface of the tape, in the form of loops between the poles.

If a tape is recorded with varying frequency but the magnetizing force is kept constant a series of different length magnets will be produced, each with the same total flux but with a concentration (flux density) that increases as the magnets get smaller, i.e. the flux per unit length of tape increases with frequency.

It is these external fields which link with the coils in the replay head as the tape is drawn past it and produce the output by induction. Being proportional to the rate of change of flux, the output increases as the flux per unit length of tape increases. The output therefore tends to be directly proportional to frequency, rising at a steady rate of 6 dB per octave over much of the frequency range. The output characteristic at each end of the usable frequency range is, however, modified by the physical characteristics of the replay head (see p. 162).

Tape–head contact

Because the field from the shorter magnets is more concentrated, it has a smaller spread than that produced by longer (lower) frequencies. The high frequencies are therefore particularly susceptible to separation between the tape and the replay head – hence the need for regular head cleaning to maintain good high frequency response. This fact also explains why cross-talk between adjacent tracks that have been recorded with insufficient spacing tends to favour the lower frequencies.

A diagramatic representation of replay head gap showing the head core and gap. The illustration shows a mid-frequency recorded on the tape at the point of a peak of a positive half cycle.

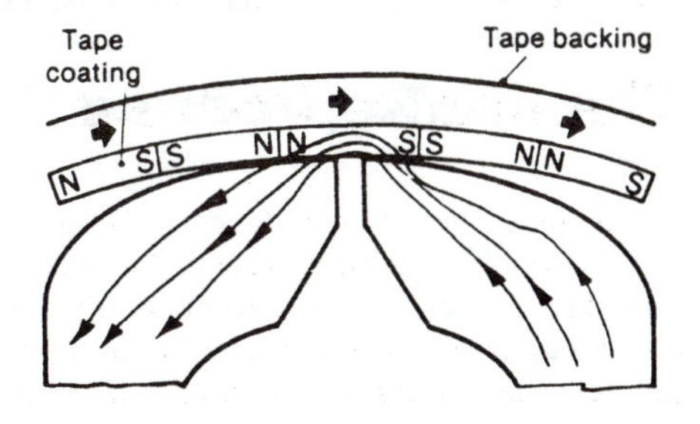

Illustrating low frequency losses due to the fact that the wavelength is longer than twice the head/tape contact area. This results in some of the flux bypassing the ring of the core.

The cancellation of output due to wavelength is equal to the gap length. (Extinction frequency.)

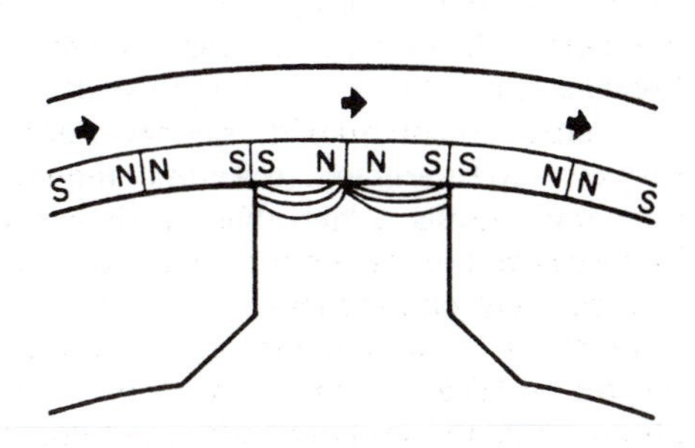

The effect of an impurity across the gap of a replay head. The magnetic flux from the tape is effectively diverted, thus reducing the output.

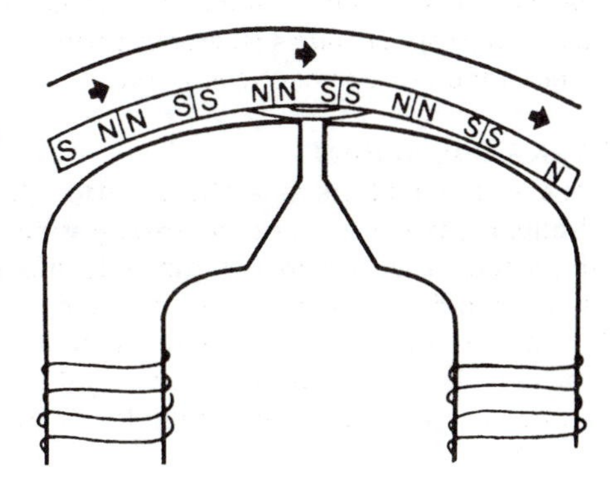

A flat overall frequency response for a magnetic record/reproduce system can only be achieved by modifying the characteristic at various points in the process.

Record/Reproduce Characteristics

As in any record/reproduce system, the object of tape recording is to produce a faithful reproduction of the original signal with nothing added, subtracted or altered. Although a straight-line frequency response is required overall, however, it cannot be maintained throughout the process because of the fundamental laws of electromagnetic induction.

Replay characteristics

The tape recorder replay head responds to the *rate of change* of the magnetic fields on the tape. If, when the recording was made, the current in the recording head was kept constant while the frequency varied, the magnetic flux induced into the tape would be constant (subject to various limitations discussed later). For most of the frequency range, however, the flux per unit length along the tape increases as the frequency increases. As the tape is drawn past the replay head, the rate of change of flux and therefore the electro-magnetic induction and the output increase in proportion to frequency. Over most of the usable range the rise is directly proportional to frequency, i.e. 6 dB per octave.

Extinction frequency

This characteristic continues up the frequency scale until a point is reached where the wavelength of the magnetic variations on the tape (which depends upon frequency and tape speed) is small enough to become comparable to the size of the replay head gap and the output falls off dramatically. When the wavelength and the head gap are the same size, both halves of the cycle occur across the gap and complete cancellation occurs. This is called the 'extinction frequency'.

Spacing loss

In practice, the replay head output does not maintain its theoretical 6 dB per octave characteristic as far up the scale as its dimensions would predict; neither does the extinction frequency occur at exact multiples of head gap size, due to some inevitable separation between the tape and the head gap. This loss is directly proportional to the separation and to the frequency.

Thickness loss

The thickness of the tape can also add a proportionate spacing loss at high frequencies because, apart from that proportion of the total flux that is present on the surface, the flux in the inner areas is spaced away from the head gap by the thickness of the tape.

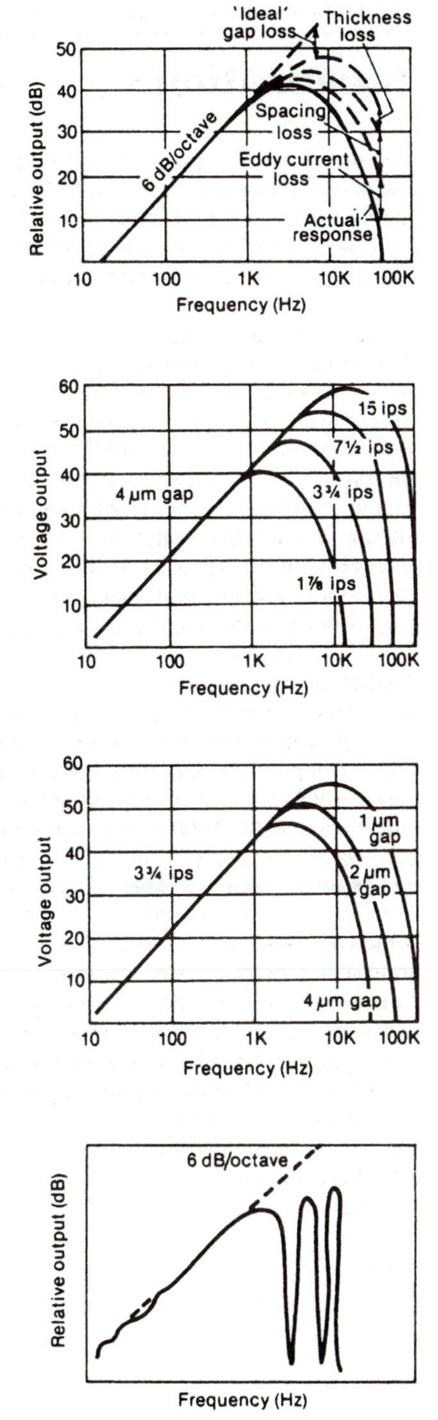

Illustrating how the basic 6 dB per octave playback curve is modified by thickness loss, spacing loss and eddy current loss.

Showing how the high frequency response varies according to the tape speed.

Showing how the high frequency response varies according to head gap size, for a speed of 3¾ ips.

Recording characteristic showing general rise of 6 dB per octave, with uneven low frequency response due to magnetic flux loss. Note that extinction frequency minima occur below the frequencies where gap size is an exact multiple of wavelength. This is due to spacing loss which effectively increases gap size.

The non-linear response of the magnetic record/replay process makes it necessary to employ equalization to both functions.

Equalization

On page 166 we showed that the output of a tape replay head has a response which rises proportionately with frequency over most of the usable range and falls off rapidly at each end of the scale. In order to reproduce the original signal faithfully it is, therefore, necessary to apply correction by altering the response of amplifiers at various points in the process.

Method of correction

The process of correcting for an uneven frequency response is called equalization. It could be applied in various ways to overcome the basic non-linearity of the magnetic recording system. For instance, one approach to the problem of increasing output with frequency would be to supply the recording head with a signal current inversely proportional to frequency over the major part of the scale; the recorded flux density would then be linear and the replay output flat. In fact, this is not practical as the tape would not accept the necessary dynamic range.

Replay correction

Most of the correction is, therefore, applied during replay. The equalization is basically the inverse of the 6 dB per octave rise inherent in the magnetic induction process, with the curve flattening off at the top end of the scale to compensate for the high frequency losses (see p. 166). The amount of high frequency boost that can be applied during replay is, however, limited by the increase in the level of tape hiss that would result.

Recording characteristic

The recording characteristic is basically flat over most of the range but sufficient boost is applied to the high frequencies in recording to make good the losses. These are due to self-demagnetization, decrease in permeability of the recording head and tape, eddy current losses, tape thickness and the erasing effect of the high frequency bias current.

In one system (NAB, described on p. 170) the low frequencies are also enhanced during recording to compensate for the low frequency sensitivity of the replay process (when the rate of change of flux across the head gap is very small), and thereby improve the programme:noise ratio.

Further equalization can also be applied to take account of different tape coatings.

Overall record/reproduce

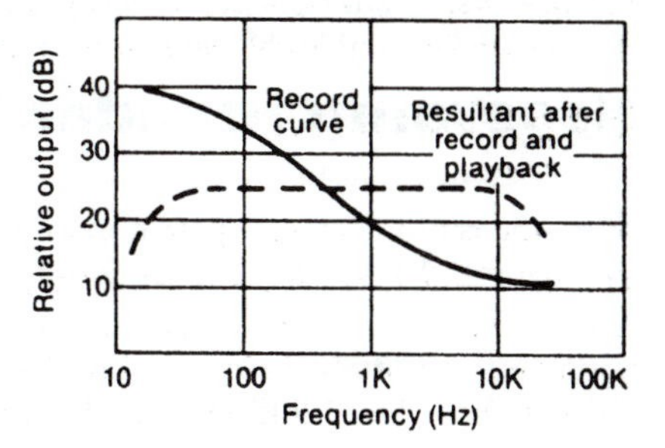

Recording and replay characteristic curves

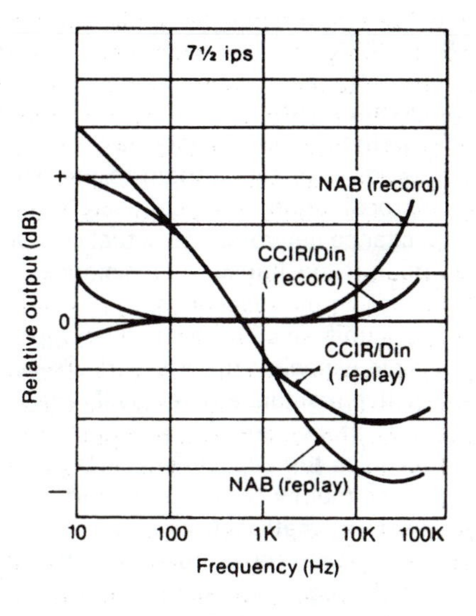

Replay curves showing differing responses for different head gap widths for speeds of 1⅞ in/s, 3¾ in/s and 7½ in/s

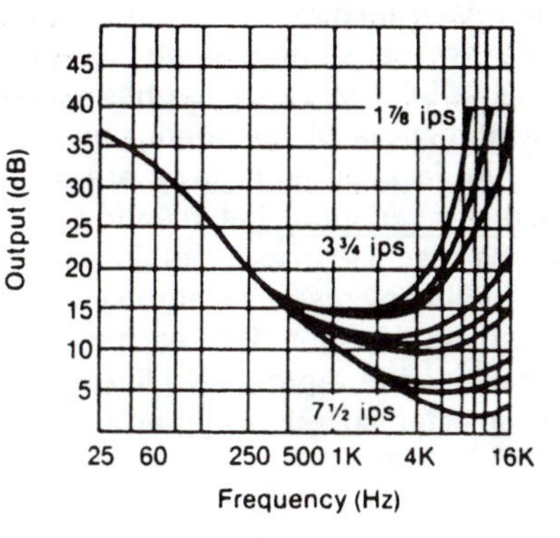

The use of standard equalization curves enables recordings made on one machine to be reproduced faithfully by another.

Recording Standards

The process of correcting for non-linear frequency response by applying the inverse characteristic is known as equalization.

Standard equalization

In the case of recording systems it is important to lay down standards of equalization so that recordings made on one machine are playable on another. The standards used are generally those developed by the National Association of Radio and Television Broadcasters NARTB (or NAB) or the DIN standards, which are modifications of the International Radio Consultative Committee (CCIR) standards by the German Standards Organization.

Time constant

The method of stipulating the prescribed curves used in the international standards is to relate them to the variation in impedance or admittance (reciprocal or impedance) with frequency of a simple resistance and capacitance network. The actual component values are not given but their product (capacitance × resistance) is quoted in the form of a time constant, in microseconds.

The significance of this will be appreciated by considering the action of a simple capacitor/resistor series network. At low frequencies, when the capacitive reactance is relatively high, the resistance can be ignored and the admittance increases almost in proportion to frequency: at high frequencies the capacitive reactance can be ignored and the impedance becomes mainly resistive and virtually independent of frequency.

The combination thus produces a curve that represents a smooth transition between a nominal 6 dB per octave slope and linear, centred about the point where the 'idealized' curves of their individual impedances would meet. The 'turn-over' frequency is the point at which the curve deviates 3 dB from the straight line.

Tape standards

Tape recorder standards are set by means of a standard recorded tape on which a full frequency run has been recorded, with such a characteristic that the curve of remanent flux on the tape versus frequency obeys a definite law. Different laws are specified for different tape speeds and there are variations between the various standards. These characteristics can be specified in terms of time constants. Where it is necessary to adjust the frequency response at the bass and treble ends of the spectrum, time constants are quoted for each turn-over frequency. Playback equalization is designed to reproduce a linear output from these curves, in relation to the selected speed, taking account of the replay head characteristics.

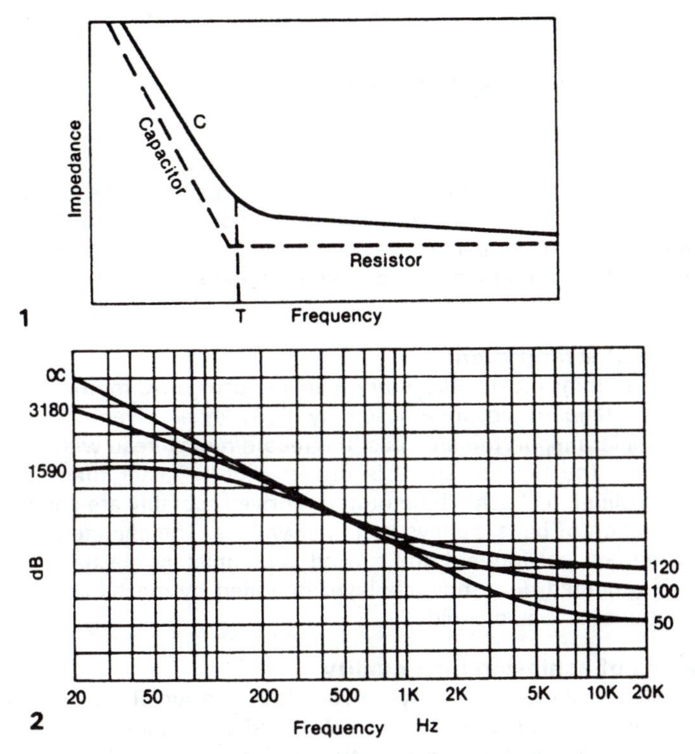

Standard tape replay characteristics and time constants

	4.75 cm/s		*9.5 cm/s*		*19 cm/s*	
Standard	*Bass*	*Treble*	*Bass*	*Treble*	*Bass*	*Treble*
CCIR	–	140	3180	120	3180	100
DIN	1590	120	3180	90	3180	50
NARTB	–	–	3180	100	3180	60

1. Curve relating impedance to frequency in a circuit consisting of a resistance and capacitor in series. The resistor line represents the impedance of the resistive element, which is independent of frequency. The capacitor line represents the impedance of the capacitor, which is inversely proportional to frequency according to the equation $Z=1/2\pi fC$, where Z=impedance, f=frequency and C=capacitance. This curve reduces at the rate of approximately 6 dB per octave. If the two impedances are connected in series the result with be a curve as in c. The 'turn-over' point or 'corner frequency' (i.e. the point on the curve nearest to where the two 'idealized' curves would meet if they continued in a straight line) occurs at $T=1/2\pi f$ where T is the time constant. The time constant (in seconds) is the product of the resistance (in ohms) and the capacitance (in farads) $T=C\times R$. In practice, because of the size of the units involved, the time constant is normally expressed in microseconds.

2. Time constants are a very convenient way of describing the shape of response curves. The graph shows theoretical tape replay characteristics with their bass and treble turn-over frequencies.

Cassette Recorders

The compact cassette recorder is effectively a miniature reel-to-reel tape system in which both spools are contained in a single plastic container thereby eliminating the cumbersome business of threading the tape. The cassette therefore combines the simplicity of operation and ease of storage of the gramophone (phonograph) record with the ability to record.

Mechanical arrangement
The normal domestic mains/battery machine uses a DC motor drive and an almost standardized drive assembly. A flywheel-driven capstan and pinch wheel transport the tape past the record/replay head with constant speed while the tape-up spool is driven slightly faster than required through a slipping clutch. The motor and drive assembly are coupled via a combination of belts, pulleys and idler wheels. The change-over from record, replay, fast forward and rewind is completely mechanical; each pulley is mounted on a movable lever operated by mechanically linked push buttons, or via solenoids.

Problems of achieving high quality
Some of the limitations of this system are: low tape speed (1⅞ in/s); narrow track width (the ⅛ in tape has four tracks); and small spools.

The low tape speed and narrow tracks give rise to problems in head design, i.e. gap size, and the tracking arrangements required to ensure good azimuth alignment and head contact. The small size of the spools and the fact that the tape guides are contained within the cassette can result in wow and flutter, which can vary from cassette to cassette and at different positions on the tape.

Near high-fidelity results are achieved in some of the better models by various improvements in cassette drive mechanisms, which are usually powered by AC motors. One type uses a bi-peripheral flywheel to drive the spool spindles. Another uses a dual capstan-pinch wheel drive, i.e. before and after the record/replay head, in order to completely isolate the tape transport past the head from the changes in tension produced by the internal mechanism of the cassette.

Further improvement, even to the domestic machine, is also possible by introducing electronic speed regulation, the control system being contained in one integrated circuit. This can almost eliminate the speed variations due to such factors as varying supply voltage, ambient temperature variation, and variations of loading of the tape to the capstan (due to tape drag in the cassette).

Digital cassette recorders
Digital cassette recorders are capable of much higher quality than is achievable with analogue machines, as they have to recognize only two states in the signal (1 or 0). See p. 194.

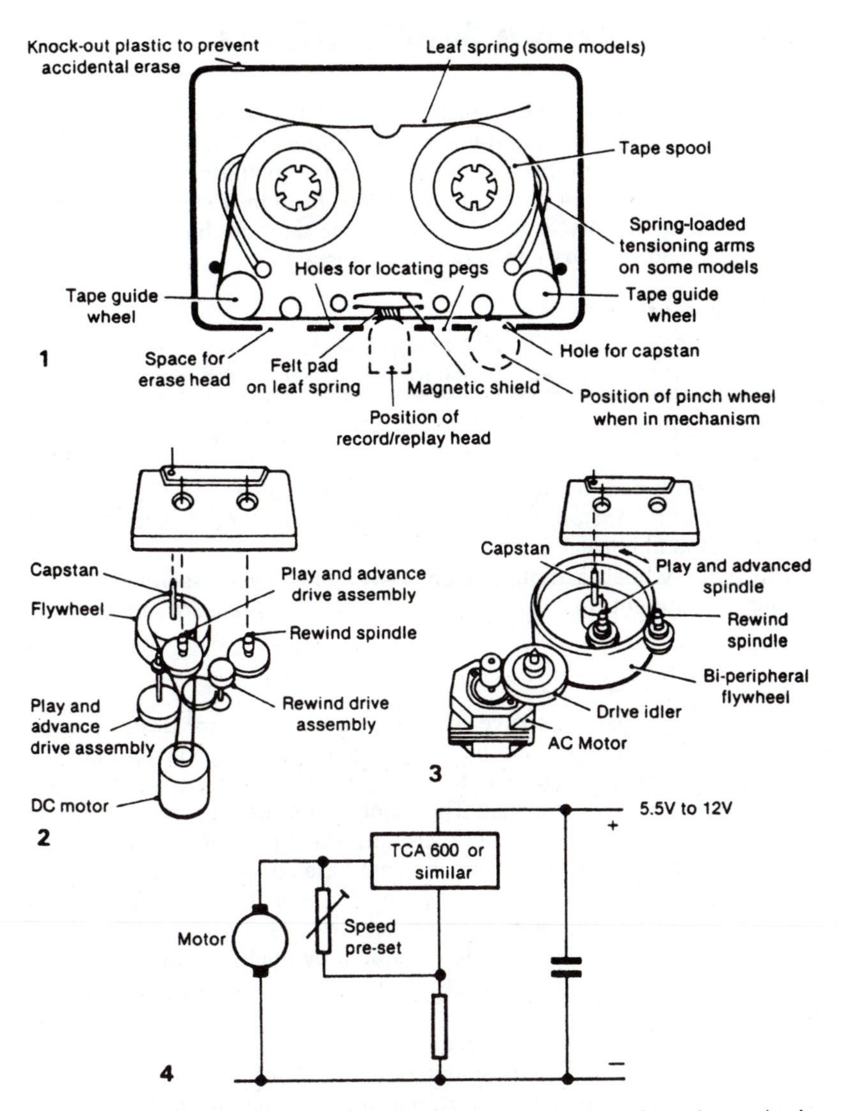

1. View of cassette with top cover removed. There is a piece of knock-out plastic or hole covered by tape to prevent accidental erasure, which is removed for recording. The tape spool is free to locate on the spindle when inserted in mechanism.
2. A typical motor drive assembly.
3. Bi-peripheral flywheel drive assembly.
4. Electronic speed regulator. The control circuit is contained in a single integrated circuit, a TCA 600 or similar. The speed of the motor is adjusted by the variable resistance preset.

Cassette drive systems are arranged to locate and couple with the spools when the cassette is placed in the machine.

Cassette Drive Mechanisms

The drive to the tape reels, which are inside the cassette, is through a spindle assembly with spring-loaded dog clutches. These automatically locate into the slots in the reels when the rotation starts.

The right-hand take-up spindle is driven through a slipping clutch to allow its speed to vary as the diameter of the tape on the reel increases. The tape speed past the heads remains constant.

Head assembly

The erase head, record/playback head, tape guides and pinch wheel are assembled on a sliding plate which moves forward in the record or replay position.

When the cassette is inserted the heads and pinch wheel penetrate it through the openings provided. This allows the heads to press the tape against the felt pad inside the cassette and the pinch wheel to press the tape against the capstan. The tape guides, which are attached to the head, hold the tape in correct alignment.

The pinch wheel is mounted on a swinging lever assembly, with a spring adjustment to give exactly the correct pressure against the capstan.

Drive mechanism

Except for the most expensive machines, a single motor drive is used. This is usually a DC motor. Some form of speed regulation is normal in all but the cheapest, most basic units. A single belt drive is used to the reels via pulleys and to the capstan. The combination of a somewhat elastic belt, rubber-tyred pulleys and the large flywheel provides the necessary 'smoothing' required to maintain constant tape speed.

The different drive conditions required for record/replay, fast forward and fast rewind are provided by physical movement of the pulleys by a system of levers and sliding plates operated by push-button controls.

In some high-quality machines, twin capstans are employed to improve the smoothness of the drive.

In the latest and most sophisticated machines, the use of logic control circuits, in conjunction with solenoid-operated transport mechanisms, enables the various functions to be selected by touch switches.

The logic system is programmed to operate in correct sequence so that if, for instance, the playback button is pressed during fast rewind, the machine will automatically come to a stop and change to playback in a smooth manner.

Some machines provide a 'search' facility where the machine can be 'programmed' to find a particular item on a recorded tape. It does this by counting the number of spaces between the items while in fast forward or rewind mode and comparing it with a preset figure.

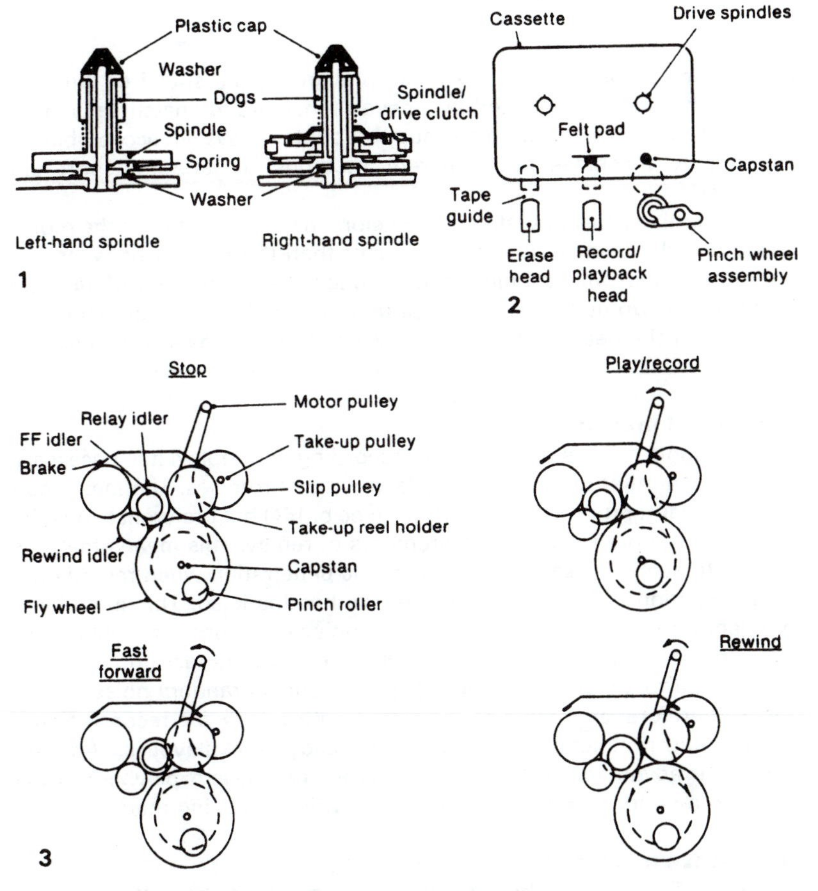

1. Cross section of a left-hand and right-hand spindle.
2. The relationship between the cassette, the tape heads and pinch wheel. The pinch wheel, record/replay head and erase head move up to the dotted positions in record/replay modes.
3. Schematic diagram of typical tape drive mechanism showing the arrangement of pulleys and idler wheels for each of the four conditions: stop, play/record, fast forward and rewind.

Continuous development has resulted in cassette machines capable of quality comparable with disc.

High-Quality Cassette Recorders

There has been continual development in the manufacture of tape, the electrical processing and in the design of the machines themselves, to the point where top-quality cassette recorders are capable of providing remarkably good quality.

Double capstan drive

With the methods of drive described previously, the speed of the tape passing the heads is affected by uneven tension and snatch caused by variable torque in the take-up and supply spools. This is not acceptable for high-quality machines and different methods of tape transport are employed.

In the double capstan system the tension on the belt between the motor pulley and the leading capstan is greater than the tension between the leading and the trailing pulley and between it and the motor. The tape between the two pulleys is thus kept taut, under slight tension, so that it passes over the heads, which are between the two capstans, at a steady speed.

Three-head machines

When a single head is used for both recording and replay, it is necessary to make a number of rather unsatisfactory compromises. The ideal head gap size is larger for record than replay (see p. 154) but preference must be given to the more stringent requirements of replay. This means that it is difficult to achieve a satisfactory magnetic penetration when recording.

The situation is made worse if a ferrite head is used. This material is excellent for recording but can be rather noisy when used for replay with a very narrow gap, as it has a porous structure with random holes which can amount to a series of dummy gaps, producing random noise.

It is therefore preferable to employ ferrite heads for recording and erasing and mumetal heads, e.g. Permaloy, for playback. Another significant advantage of separate heads for record and playback is the ability to monitor a recording of the tape while it is being made.

Servo-drive machines

It is very difficult to maintain constant tape speed with cassettes because of the varying friction of the pressure pad and imprecise guide rollers. Even slight variations will create wow and flutter and affect the frequency response by altering the tape tension.

These problems are overcome in high-quality machines by controlling the tape speed by an electronic logic servo system driving the motor, controlled by a quartz-locked oscillator.

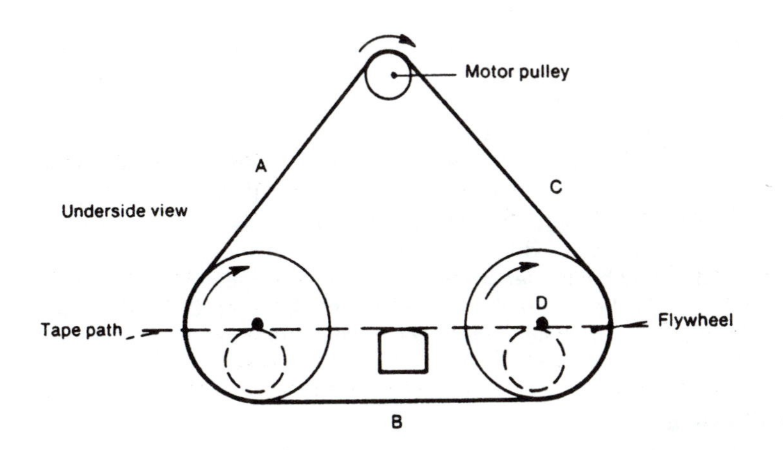

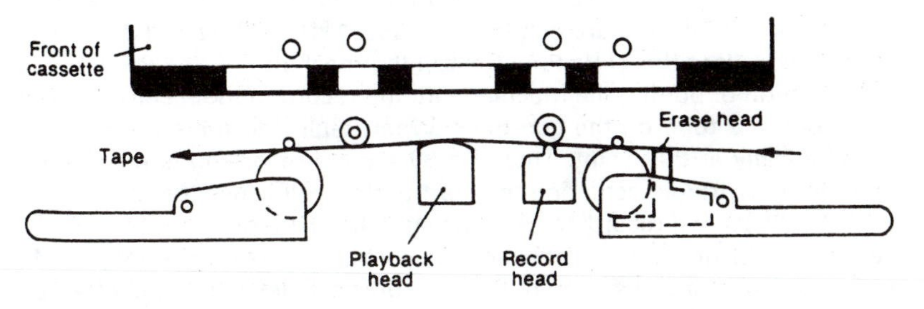

1. Underside view of cassette drive mechanism showing belt drive arrangement. The belt is taut at *A*, slacker at *B* and quite slack at *C*. The tape is therefore kept taut between the capstans (*D*).
2. General arrangement for a three-head twin-capstan machine, showing relationship of the heads to the cassette openings. High-quality tape machines have separate record and replay heads so that each can have its optimum gap width – typically 5 μm for record and 0.7 μm for playback. It also enables simultaneous playback during recording for checking.

NAB cartridge machines provide the facility of rapid selection and instantaneous cue.

Broadcast Cartridge Machines

Broadcast cartridge machines, employing the NAB format, are intended to provide a source of programme material that can be quickly selected and accurately cued. It is the ideal medium for inserting short recordings such as announcements, jingles, commercials and sound effects etc. Cartridges are sometimes used to record short sequences for post-dubbing to film.

Continuous loop cartridges

The cartridges contain a continuous loop of lubricated tape. This is pulled out from the middle of a single spool and returns to its periphery via guide pulleys, set in the corners of the plastic container.

Pressure pads are provided behind the tape at the points where the record and replay heads penetrate when the cartridge is engaged. There are three standard cartridge sizes, designated A, B and C. Spool lengths can vary from about 20 seconds to 31 minutes, with the tape running at 3¾ in/s.

Cueing system

Unlike the eight-track cartridge system, which uses metal foil cueing, the NAB system uses audio frequency tones, recorded on a separate track, to start and stop the machine and perform additional functions if required.

There are three standard cue tones in use: 1 kHz, 150 Hz and 8 kHz. The basic action uses the 1 kHz tone to stop the machine.

The action of putting the machine into the record mode records a short burst of 1 kHz tone on the cue track. When replaying the tape will stop automatically at this point. Thus if a 30-second recording is made on a 40-second cartridge, actuating the start button will cause the machine to play the 30-second recording from the start. It will then run on a further 10 seconds with no output until the cue comes round on the loop at 40 seconds, where it will stop with the tape precisely located to play the cue again.

External cueing

Where machines are equipped with the two additional tones, the 150 Hz (auxiliary) tone is usually used as an end-of-message signal which can be used to start the next machine in a sequence. The 8 kHz (auxiliary) tone can be used for extra functions such as cueing superimposition of effects or picture slide changes etc.

Machine capability

Professional cartridge machines are capable of quite good quality. They are normally capable of mono or stereo recording and playback. Some have fast forward facility (two to three and a half times play) to expedite cueing. It is not normal to provide erase facilities.

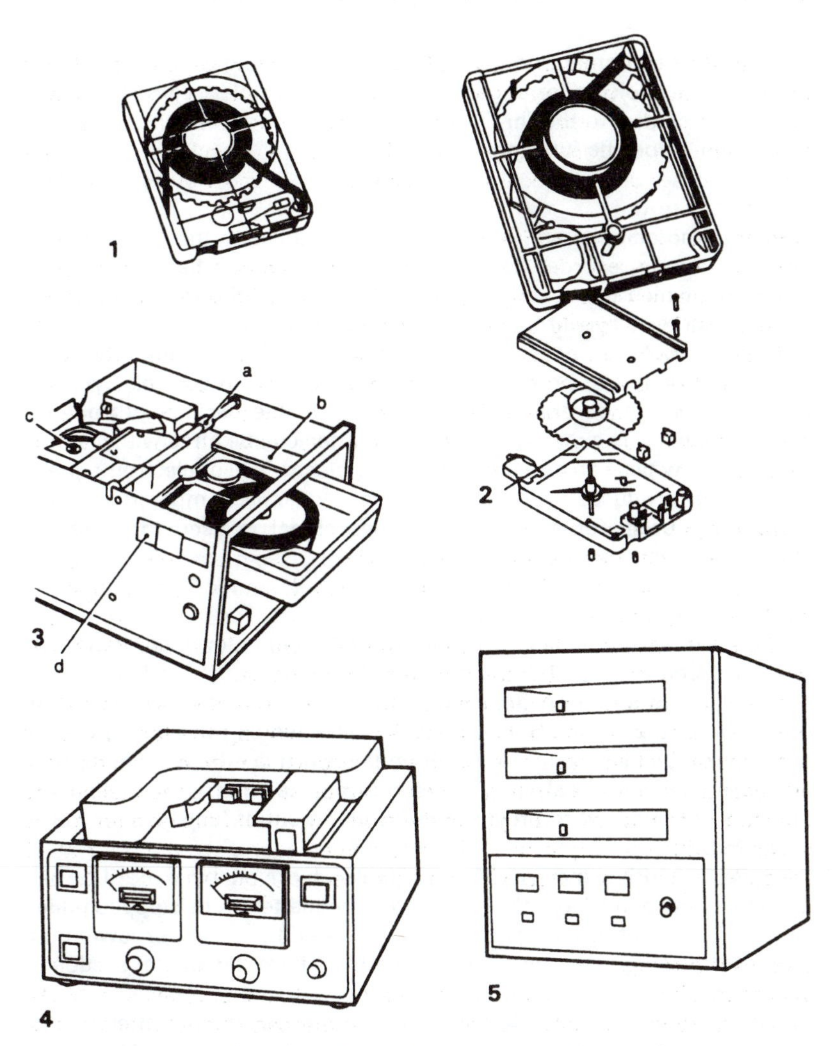

1 and 2. Two types of cartridge. Note the difference in the position that the tape feeds from the spool.
3. Cartridge machine mechanism. *a.* Spring-loaded hold-down rollers. *b.* Tapered cartridge guide. *c.* Adjusting screws. *d.* Cartridge loading spring.
4. General view of cartridge machine showing position of the heads. This is a single machine.
5. Machines are available with up to three cartridge slots stacked in vertical formation.

High-quality disc reproduction begins with a good turntable.

Vinyl Disc Reproduction

The range of equipment now available for the playing of disc recordings offers a wide choice of arrangements from the complete compendium of the 'music centre' to the 'hi-fi' arrangements of independent units each chosen for a specific function.

The motor unit

Turntable motor units are usually of a type similar to the capstan-drive motor in a tape recorder (see p. 174). These units can be: synchronous motors, in which the speed of rotation is locked to the frequency of the mains; hysteresis-synchronous, which incorporate a hard-steel shell in the rotor to provide a measure of permanent magnetism and therefore stronger 'lock-up'; or brushless DC motors with servo control. Synchronous motors can be driven directly from a stable oscillator through a power amplifier. The motor, or turntable platter, can be fitted with a strobe tachometer, which can be a light and photo-electric cell or electromagnetic sensor. This produces a reference frequency which is compared with that of a stable oscillator and any difference signal is used to make the necessary speed correction.

Turntable drive

There are three methods of driving the turntable platter in common use: rim drive, belt drive and direct drive. In the rim drive method, a stepped boss transmits a friction drive via a rubber-rimmed idler wheel to the inside rim of the platter. Speed change is effected by raising or lowering the boss, or the idler, so that the idler engages with a different diameter on the boss. A mechanical linkage disengages the idler wheel when the motor is switched off to prevent the development of 'flats' which would cause flutter.

In the belt drive method, the turntable is driven by a flexible belt stretched between a large flywheel under the platter and a stepped pulley driven by the motor. The speed can be changed by raising or lowering a 'fork' which causes the belt to engage with a different diameter section of the pulley.

In direct drive, a multi-pole slow-speed motor drives the platter spindle directly. Speed can be changed by altering the frequency of the AC supply or switching the number of poles in the motor.

Rumble

An important parameter in the choice of turntables is freedom from rumble, i.e. low frequency noise transmitted from the motor to the pick-up, through either the platter or the tracking arm pivot. Stereo pickups are sensitive to vertical vibration. The level of rumble should be at least 45 dB below the signal level. This can be achieved with well-designed and independently suspended motors and drive systems.

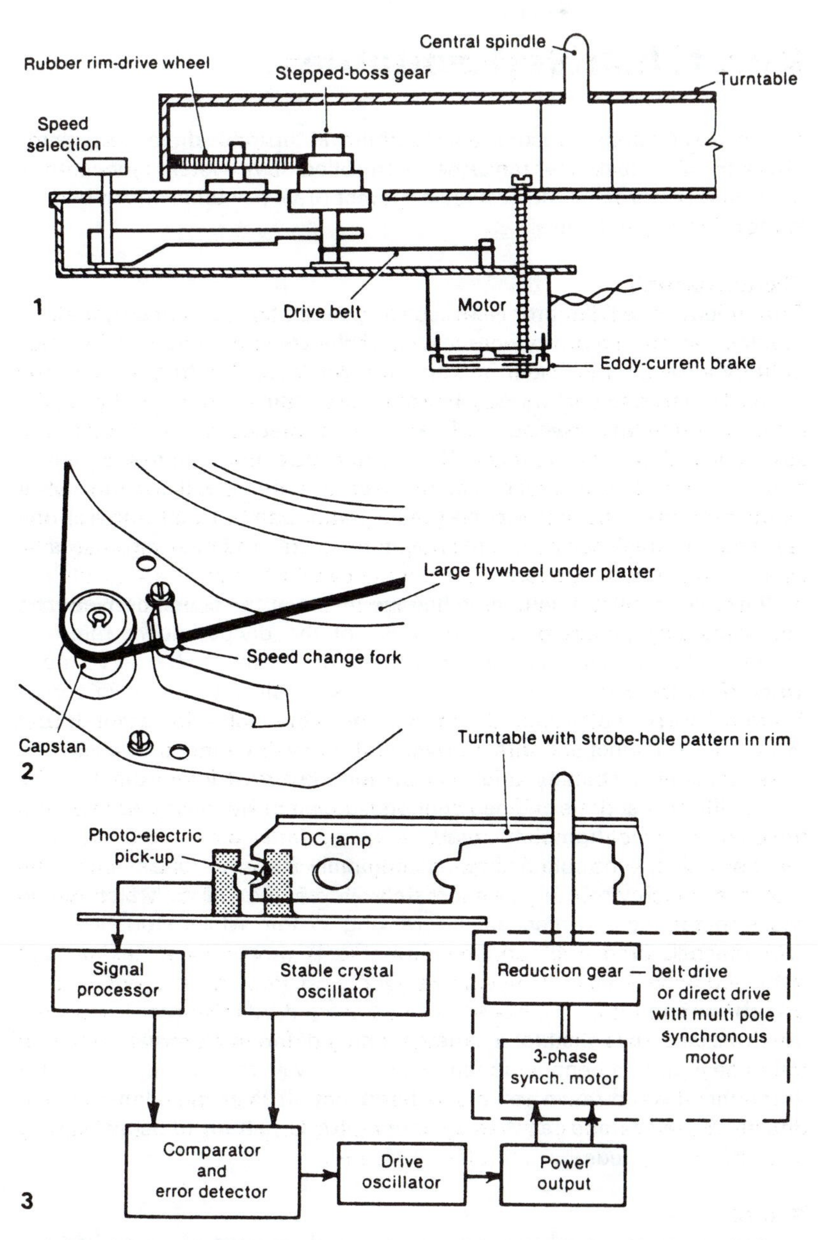

1. Outline of a four-speed deck.
2. Belt-driven turntable arrangement. The speed change fork moves the belt to different diameters on the capstan.
3. Block outline diagram of self-controlled power oscillator system.

Speed stability and high starting torque can be achieved by a direct drive motor controlled by a quartz oscillator.

Crystal Drive Turntable

The search for speed, accuracy and stability in turntable drive systems has led to the introduction of the crystal-controlled drive system pioneered by Technics. The basis of the system is a direct-drive turntable driven by a DC motor with servo control.

Speed control

The speed of the DC motor is locked to an oscillator controlled by a quartz crystal. The oscillator produces a very stable reference frequency which is split by a frequency divider into an appropriate control frequency for the selected turntable speed (33⅓, 45 or 78.26 rev/min).

Integral with the drive motor is a frequency generator which produces an AC signal related in frequency to the actual turntable speed. This is read by speed and phase control circuits. The speed control circuit converts the output of the frequency generator into the appropriate electrical voltage required to maintain correct speed regardless of changing load conditions.

The phase control circuit matches the phase of the signal derived from the frequency generator to the phase of the divided-down reference signal. Differences in phase are translated into current to the servo motor in either a forward or reverse direction as required to maintain correct speed. The phase-locking action provides very fast correction torque which, in conjunction with a relatively heavy platter, produces very accurate speed stability and almost complete freedom from wow or flutter. The direct drive and heavy platter, coupled with vibration damping, result in very low figures for rumble.

This type of turntable has an exceptionally fast start, attaining the full speed of 33⅓ rev/min in approximately 25° of revolution. It also has an electromagnetic and mechanical breaking system which stops it almost instantaneously.

Stroboscope

A strobe pattern is included on the periphery of the platter which is viewed with the light of a neon lamp fed (via special waveform-shaping circuitry) from the divided-down reference frequency. It is thus independent of mains variations and requires only one strobe pattern to accommodate the different speeds.

Pitch control

The majority of turntables provide a fine adjustment of speed within a range of about 10%. In better-quality quartz-controlled systems this is achieved by altering the frequency of the oscillator.

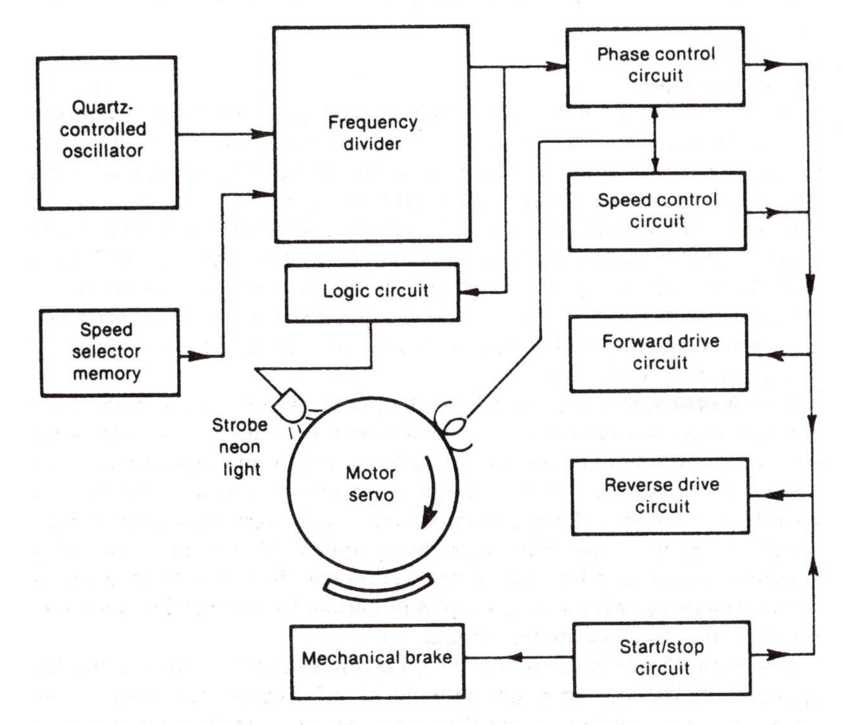

Block diagram of a crystal-controlled turntable drive system.

The frequency divider divides the oscillator frequency down to a suitable control frequency for the selected speed. The speed selector memory stores the selected speed and dictates the appropriate division to the frequency divider. The phase control and speed control circuits compare the frequency and phase of the divided control frequency with that derived from the frequency generator on the DC motor (FG). The resulting control signals are interpreted by the forward and reverse drive circuits and applied to the motor servo. A strobe neon lamp has a waveform specially shaped by the logic circuit of the divided reference frequency. Mechanical and electrical braking is applied by the start/stop circuit and the mechanical brake.

Reproducing Styli

The excessive weight of early pick-up heads and the rapid wear suffered by steel or 'thorn' styli caused serious groove wear. The life expectancy of the early 78 rev/min disc was very short.

Stylus shape

In order to obtain more faithful reproduction, modern pick-up styli are designed to complement the shape of the recorded groove. The shoulders of the stylus must track the sides of the groove, and the point must penetrate but not reach the bottom of the valley. To achieve this, the radius of the point is made about four times greater than that of the recording stylus. Enough tracking force must be applied to maintain full contact with the groove walls at all times whilst at the same time causing only negligible wear. It must be hard enough stand up to the mechanical forces to which it is subjected during use and be ground to a very smooth finish to keep contact friction low.

Stylus materials

Various materials are used for modern styli and are usually of a hard crystalline material. High-quality steel is no longer common; sapphire is most popular, with ruby or diamond being used in the more expensive cartridges. Although of hard materials, styli are liable to be chipped by rough handling and can then cause severe damage to the soft material of the modern microgroove disc. They have long working lives, ranging from several hundreds of playings for sapphire to several thousands for diamond before noticeable wear. The amount of material used in most modern cartridge styli does not allow for regrinding but styli in the form of a tipped shaft or needle can be reground for further use.

Trailing angle

The recorded groove is cut with a vertical stylus but, when replayed, optimum tracking occurs if the pick-up stylus trails at a slight angle from the vertical, usually about 10° to 15°. If the stylus is of circular cross-section this effectively increases its diameter longitudinal to the groove and results in some high frequency attenuation. This can be overcome by giving the stylus an elliptical cross-section at the point of contact, making its longitudinal thickness approximately proportional to the sine of the trailing angle. The tracking achieved with the elliptical stylus is very accurate.

High frequency attenuation due to tracking error is progressive towards the centre of the disc but this is compensated for during recording and does not require correction.

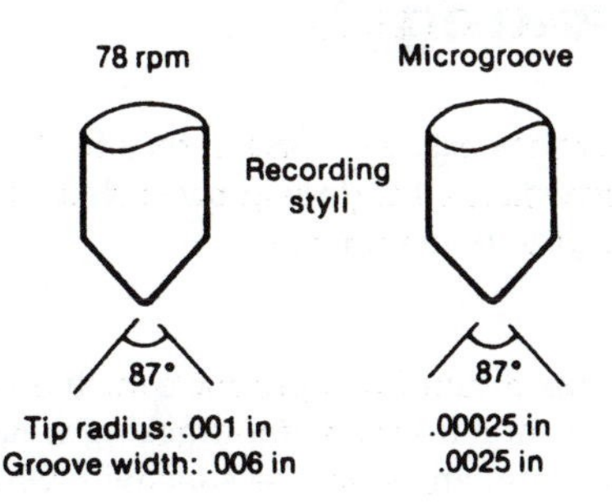

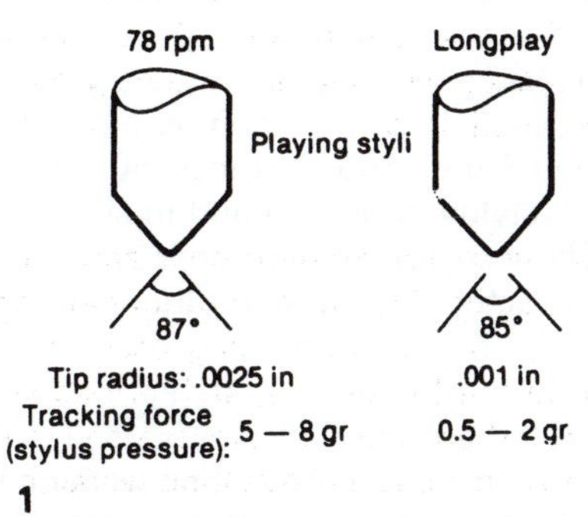

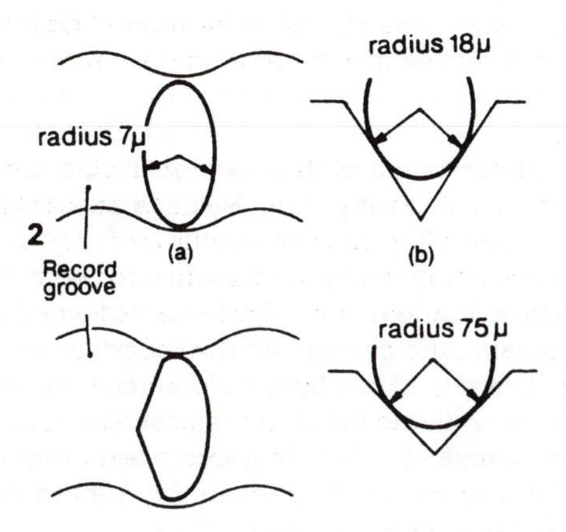

1. Comparison of recording and replay styli.
2. The replay stylus with trailing angle 10° to 15° and elliptical stylus with longitudinal cross-section elliptical (a) and contact cross-section circular (b).

Stylus Tracking

The modern microgroove disc requires a very high standard of performance by the pickup head and this depends very largely on the ability of the stylus to follow the groove variations.

Trackability
To obtain a good high frequency response with minimum wear to the stylus or record groove the tip mass and tracking weight must be as small as possible and the compliance (i.e. the ease with which the stylus can be moved from side to side) as high as possible.

On the other hand the stylus weight must be sufficient to enable it to stay firmly in the groove while being subjected to the very rapid accelerations involved in reproducing high frequencies and transients.

Tracking distortion
Any failure of the stylus to follow the contours of the groove will either result in deformation of the grooves (possibly resulting in permanent damage) or cause the stylus to ride up and in severe cases skate across the record. In either case tracking distortion will occur, which produces a 'roughness' in the character of the sound due to the addition of spurious harmonics.

Bias compensation
A possible cause of tracking distortion is the mechanical bias imparted to the pickup due to the angle of the arm with respect to the groove. Because the pickup arm is not directly in line with the groove, the drag of the stylus in the groove creates a force which acts to one side, tending to pull the arm towards the centre of the record.

To compensate for this most tracking arms are provided with 'Bias Compensation' (sometimes called 'Antiskate'). This usually takes the form of a weight suspended on a string, a magnetic bias or specially angled tension springs. The sideways 'bias' is made adjustable and should be increased in relation to the tracking weight.

Bias can be assessed using a test record with tones of various frequencies recorded at very high levels and adjusted for the minimum required to prevent audible harmonic distortion (impurity of tone). Another type of test record has bursts of filtered tone which are adjusted for visual symmetry with the aid of an oscilloscope. The well-known Shure 'Audio Obstacle Course ERA 111' provides a test using musical excerpts, recorded at five increasing levels, for the subjective assessment of trackability under dynamic conditions.

Trackability also involves the ability of the pickup to ride the up and down motion of a warped disc. This is a function of the low frequency resonance of the pickup/tracking-arm combination.

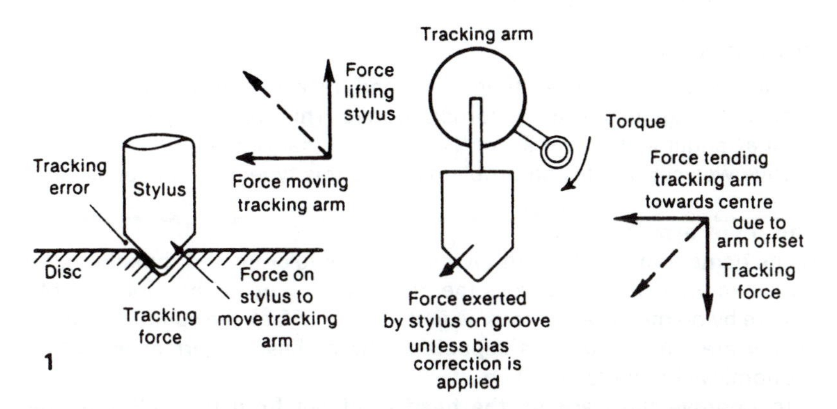

1. Tracking error. Unless the arm is dynamically balanced, part of the groove force is used to move the tracking arm which can result in tracking error. With the tracking arm dynamically balanced, the groove 'guides' the arm towards the disc centre.

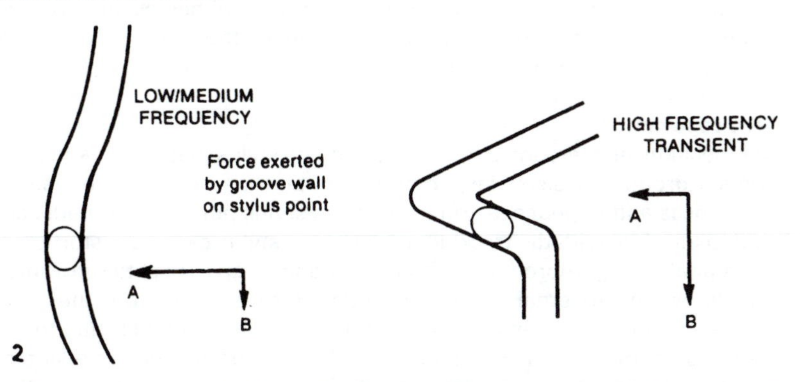

2. The replay stylus with 10° to 15° and eliptical stylus with longitudinal eliptical (A) and contact cross-section circular (B).

The recent vogue among rap musicians to rotate the turntable backwards and forwards by hand in time with the rhythm can be ruinous to the stylus.

The Tracking Arm

The purpose of the tracking arm (or tone arm) is to support the pick-up head as it tracks the groove across the disc. Two basic types are in use: the radial arm and the pivot arm.

The radial arm
The arm and head form a straight line and track linearly so that the head is always tangential to the recorded groove. This would appear to be an ideal arrangement as the action is similar to that of the cutter when the groove was cut but design problems make this type of arm expensive.

The pivot arm
As its name implies, the arm pivots about its fulcrum and thus the head travels in a shallow arc across the disc. Tracking errors are unavoidable but are by no means as serious as first appear. In fact, the compromises in design are small and arms capable of the highest performance can be economically manufactured.

To improve the tracking, the head is off-set from the radius of the pivot to be tangential to the groove at an optimum point, which is empirically determined to be 6 cm from the centre of the disc. The arm and head are balanced horizontally about the pivot and then loaded to give the required stylus pressure in the groove, either by gravity (off-balancing) or by spring loading. The action of the fulcrum and pivot must have minimum resistance.

Arm resonance
The mechanical interconnection of the groove, stylus, head, arm and pivot forms a group of resistances, compliances and masses which equate to resonances and impedances referred to the stylus point. The impedances tend to interfere with the tracking action of the stylus causing distortion of the signal being reproduced. The resonances give rise to unwanted signals. Other external effects include surface ripple on the disc, misplacement of disc centre and turntable vibration (rumble) due to the drive system or motor, cabinet, floor etc. A judicious application of choice of materials, balancing and damping during the design and construction can reduce these effects to a minimum. Resonances are constrained wherever possible to occur outside the range of the frequency band being used.

Arm–head relationship
Generally, tracking arms are designed for use either with an integral head or with replaceable heads from the same manufacturer. Some arms are designed for general-purpose use with adjustable fulcrum, pivot, balance and torque.

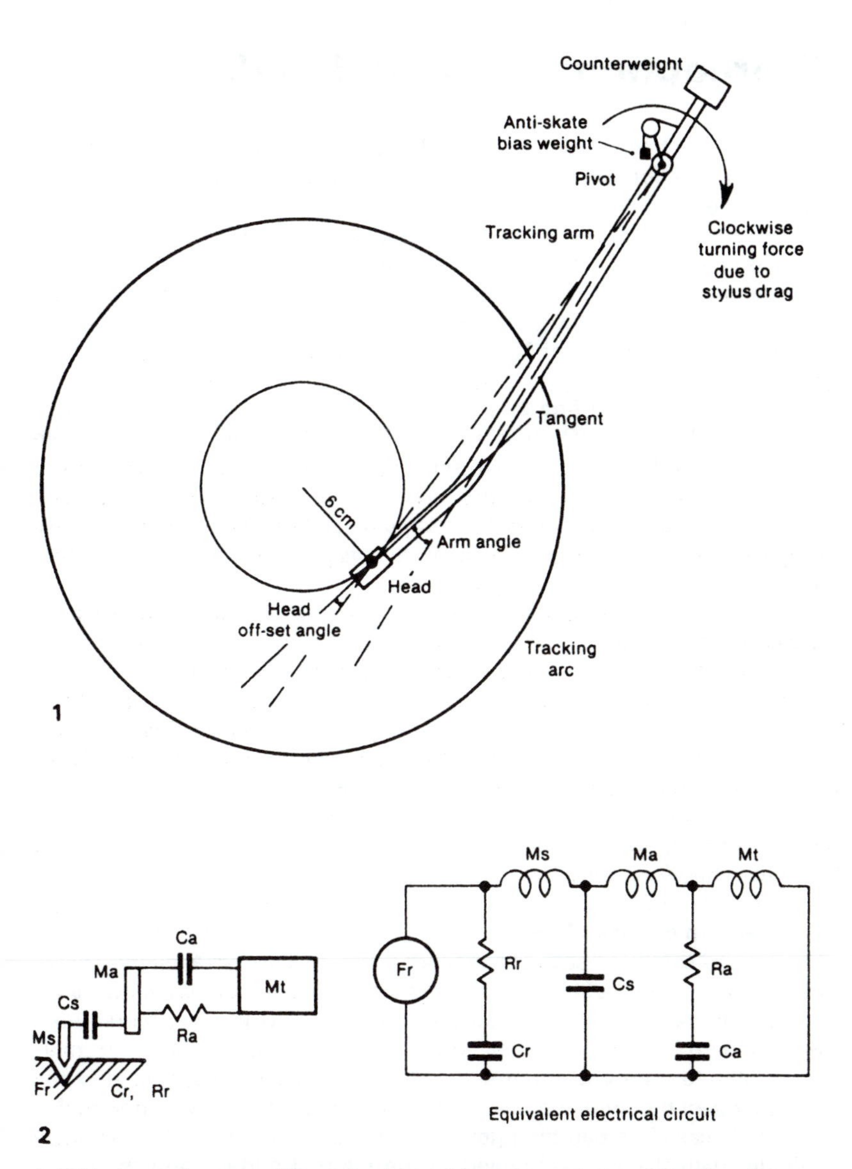

1. With the optimum tracking point located 6 cm from centre of the disc, the effect of the arm angle is to reduce the error from the agent along the tracking arc. Off-setting the head introduces torque at the stylus point which needs counter-balancing.
2. Basic head-arm circuit and equivalent electrical circuit. *Ms* = stylus mass; *Cr* = groove compliance; *Rr* = groove resistance; *Cs* = stylus–transducer compliance; *Ma* = transducer mass; *Ca* = mounting compliance; *Ra* = mounting resistance; *Mt* = tone arm mass.

Stereophonic pick-ups have to resolve groove displacement in two directions at 90° to each other.

Stereophonic Pick-up Heads

The groove is tracked by a single stylus coupled to a resolving system so that the excitation from each wall is individually conveyed to a separate pick-up element.

In stereophony, the effect of crosstalk between the outputs is a reduction of the stereo effect. The track is recorded with a channel separation of 30 to 40 dB. It is inevitable that some crosstalk will occur in the pick-up head but a good-quality head will give an output with a separation of the order of 20 dB.

Variable reluctance heads

The principle of operation is similar to that for monophonic variable reluctance heads except that two sets of poles are used set at right angles to each other. Action of the armature along one axis activates only one circuit and vice-versa. Thus the outputs generated are more or less independent of each other. Heads of this type have been developed to a very high standard and are widely used.

Moving-coil heads

The stylus is compliantly coupled to two small coils set at ±45° from the vertical in a single longitudinal magnetic field. The coil assembly pivots on a very precise centre axis; the armature couples at 90° to this axis such that the action of the armature at (say) +45° causes one coil to cut across the magnetic field generating an output whilst the other rotates around the lines of force with no effect. Thus each coil generates an independent output. For best results, the coil assembly needs to have small mass; the outputs, therefore, tend to be low level.

Crystal and ceramic heads

Crystal and ceramic heads consist, generally, of two crystal elements in a ceramic mounting at 90° to each other. They are coupled to the stylus via a parallelogram of compliant levers arranged so as to give the required flexing to each element. One bar of the lever system is fixed and the stylus is coupled to the opposite arm. The two elements are connected to the mid-points of the remaining two arms. The whole assembly is mounted in a ceramic case in a damping gel. Another type is in the form of a hollow cylinder with the stylus coupled to the internal surface and the transducers to the external.

Semiconductor heads

Because of the wide bandwidths possible with semiconductor heads, they have proved to be eminently suitable for use with the single groove stereo-quadraphonic disc.

Standard colour code for stereo pick-up leads

No. of leads	*Left Channel*	*Right Channel*	*Ground*
3	White	Red	Black (common)
4	White, Blue*	Red, Green*	–
5	White, Blue*	Red, Green*	Black

*Blue and Green are the 'earth' ends

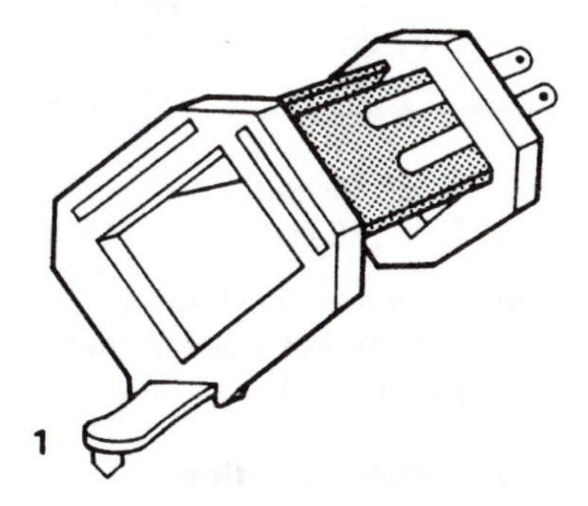

1 Basic construction of a crystal stereophonic pick-up.

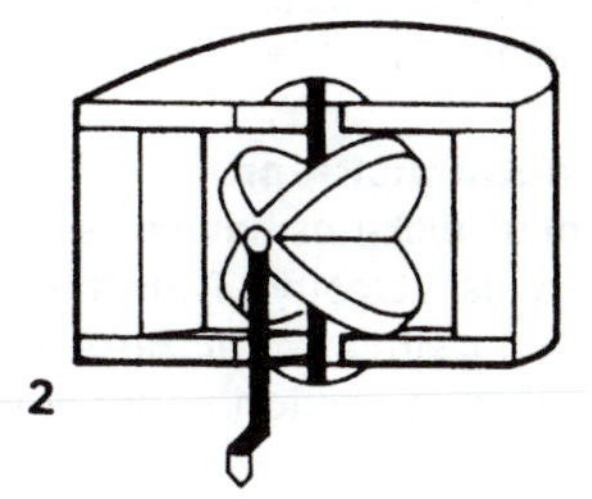

2 Basic moving-coil stereo head showing the centre balanced cross-coil assembly with the armature coupled to the horizontal axis.

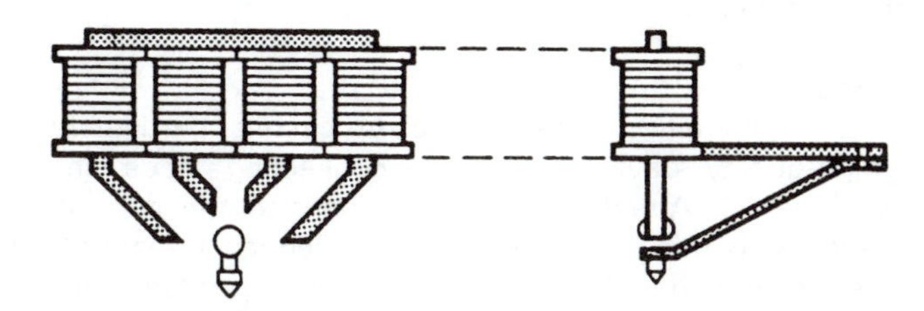

3 Outline of a variable reluctance stereo head. The side elevation shows the compliant arm supporting the magnetic armature.

Pick-up Characteristics

During recording the groove deviation is restricted at low frequencies and boosted at high frequencies. This is done according to a standard laid down by the Record Industry Association of America and known as the RIAA characteristic. For a constant recording level this gives a constant groove deviation below 600 Hz and a groove (stylus) velocity proportional to frequency above 600 Hz.

This recording characteristic is used by most commercial disc manufacturers modified as being the combination of three separate curves defined by time constants, these being:

	Fine groove	*Coarse groove*
Treble:	75 μs	50 μs
Middle:	318 μs	450 μs
Bass:	3180 μs	3180 μs

An inherent advantage is the improvement in the signal:noise ratio since disc surface noise increases with frequency due to the structure of the material used.

Radius compensation

During recording high-frequency boost is also used to compensate for stylus tracking errors due to the fact that the groove longitudinal speed decreases towards the centre of the disc. This is in addition to the RIAA characteristic.

Magnetic pick-ups

Magnetic pick-ups have an electrical output approximately proportional to the stylus velocity. Thus the pick-up output will closely resemble the recording signal after RIAA compensation and will need to be passed through a correction network with an inverse characteristic.

Crystal pick-ups

Crystal pick-ups, including semiconductor heads, have an output proportional to groove deviation. They will, therefore, have an output signal tolerably flat in response and do not require any correction if terminated in the correct impedance.

Connecting leads

High gains are generally involved in pick-up circuits, particularly in the case of high-quality devices, and the connecting leads should be effectively screened. Attention should be paid to the capacitance of connecting leads as this may affect the high-frequency response of the head. Some cartridge manufacturers assume lead capacitance in the head design; thus very low capacitance leads will affect their response.

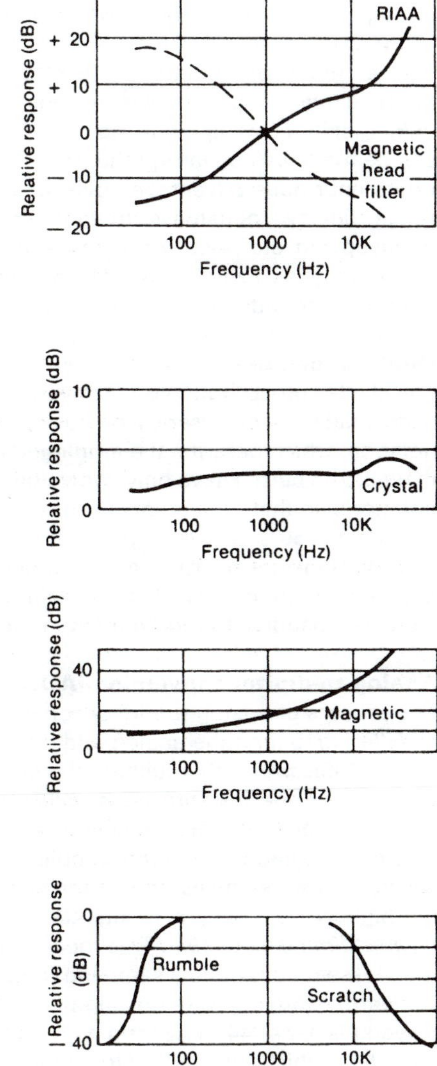

Relative response showing the RIAA recording characteristic and the inverse RIAA filter response required for magnetic heads.

Typical response of a crystal pick-up.

Pick-up head output from RIAA Standard Test Disc.

Typical rumble and scratch filter responses.

Digital techniques offer considerable improvement in audio quality over analog systems.

Digital Audio

The technology for processing audio by analog means (in which the strength and shape of the recorded waveform bears a direct relationship to the original sound waves) has been extended to the point where the 'law of diminishing returns' operates. Few further improvements are possible and would involve a considerable increase in the cost and vulnerability of equipment. However, a very considerable improvement can be made by converting the analog audio signal into digital format. It offers enormous advantages over analog systems in terms of dynamic range, fidelity, robustness, freedom from degradation in recording and multicopying etc. Also, once the audio signal is in digital format, the opportunity is opened up to use computer techniques to manipulate and record the sound.

Digital techniques

The digital process involves converting the analog audio signal into a code made up of a series of pulses, representing numbers in binary notation, which describe the amplitude of the waveform at each moment in time. The pulses have only two values, high and low (1 and 0 in positive notation), so that the equipment has only to recognize two states: above and below 50% of the high value. This means that material can be recorded and re-recorded any number of times without suffering any degradation in quality. This is achieved at the expense of a vastly increased bandwidth requirement in the equipment.

Analog-to-digital conversion (ADC)

The process of analog-to-digital conversion consists of:

1. *Filtering* to limit the analog audio input bandwidth to prevent *aliasing* (see p. 198), while still retaining the desired audio frequency response.

2. *Sampling* the instantaneous values of the waveform at a fixed rate. Obviously the higher the sampling rate the better, as the analog signal is varying continuously and the sampling process ignores any changes that occur between samples. In fact the sampling rate must be at least twice the highest frequency it is intended to convert. This is because of the *Nyquist limit* (see p. 198). This means that a full audio spectrum of 20 Hz to 20 kHz would require a sampling frequency of at least 40 kHz. In fact the standard sampling rate for domestic audio equipment (such as CD players) is set at 44.1 kHz (originally chosen to be compatible with video recorders, which can be used for recording digital audio). The standard for broadcast transmission (where the audio range can be limited to 15 kHz) is 32 kHz and for digital tape recording is 48 kHz (for high quality and to allow some scope for speed variation).

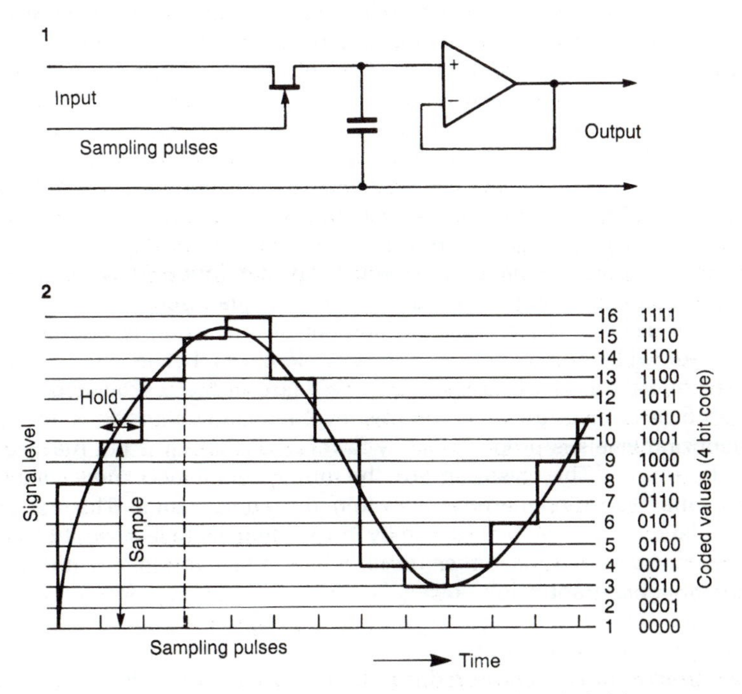

1. The basics of a sample-and-hold circuit. The electronic switch (shown as an FET) opens and closes at the sampling rate, imposing a charge on the capacitor, in proportion to the instantaneous value of the audio signal at each moment in time. The capacitor holds the charge while it is measured and until the switch closes with the next sample.

2. Sampling an analog waveform. For simplicity only a four-bit code is shown, which would allow only 16 levels of measurement (quantizing levels). This would result in a high degree of error and thereby a high degree of quantizing noise. The 16-bit code (as standard for compact discs) produces 65 536 sampling levels with a signal-to-quantizing noise level in excess of 90 dB. In the reverse process (digital-to-analog conversion) the digitized signal is converted to a series of levels corresponding to the original samples, presented at the same pulse (clocking) rate as the original. This produces a similar, but jagged, waveform which, after passing through a low-pass filter, resembles the original analog waveform.

Samples of the audio waveform have to be converted into a digital code.

Analog-to-digital Conversion

On page 194 it was explained that samples of the analog audio signal are taken at a fixed rate. These are held in a store while they are measured and converted into a code. In its simplest form the store can take the form of a capacitor which is charged through an electronic switch that closes and opens at the sampling rate.

Quantization

The process of measuring each sample involves comparing it against a scale consisting of a number of discrete values called *quantizing levels*, each of which is assigned a reference number to represent its value. As the analog signal can have values which vary continuously over the whole range of measurement the actual level and the quantized level will seldom coincide exactly. The greater the number of quantizing levels the more accurate will be the measurements. Differences between the sample levels and the quantizing levels can be heard in the reproduced sound. The noise is most apparent on low-level signals where the number of quantizing levels is proportionately fewer and where it is not masked by louder sounds. The noise can take the form of *granulation noise* due to a beat effect creating harmonics within the audio range. The effect of granulation noise can be reduced by introducing a small amount of audio white noise known as *dither noise*, which effectively masks the more objectionable granulation noise.

Non-linear quantization

Quantization noise can be reduced by a process of non-linear quantization, in which the number of quantization steps is increased as the level is reduced (i.e. where it is more critical) in the ADC process and increased proportionately in digital-to-analog conversion. This is a form of companding, analogous to Dolby and other audio NR systems. It is not appropriate for use in signal processing systems, where adding and fading of signals is involved.

Coding

The quantized levels are converted into a code, usually binary, which represents them as a group of digits composed of high or low signals. A binary code with 10 digits will provide 1024 (i.e. 2^{10}) quantizing levels, with a signal-noise level of 60 dB. The 16-bit code, as standard for compact discs, produces 65 536 levels, with a signal-to-quantizing noise level in excess of 90 dB.

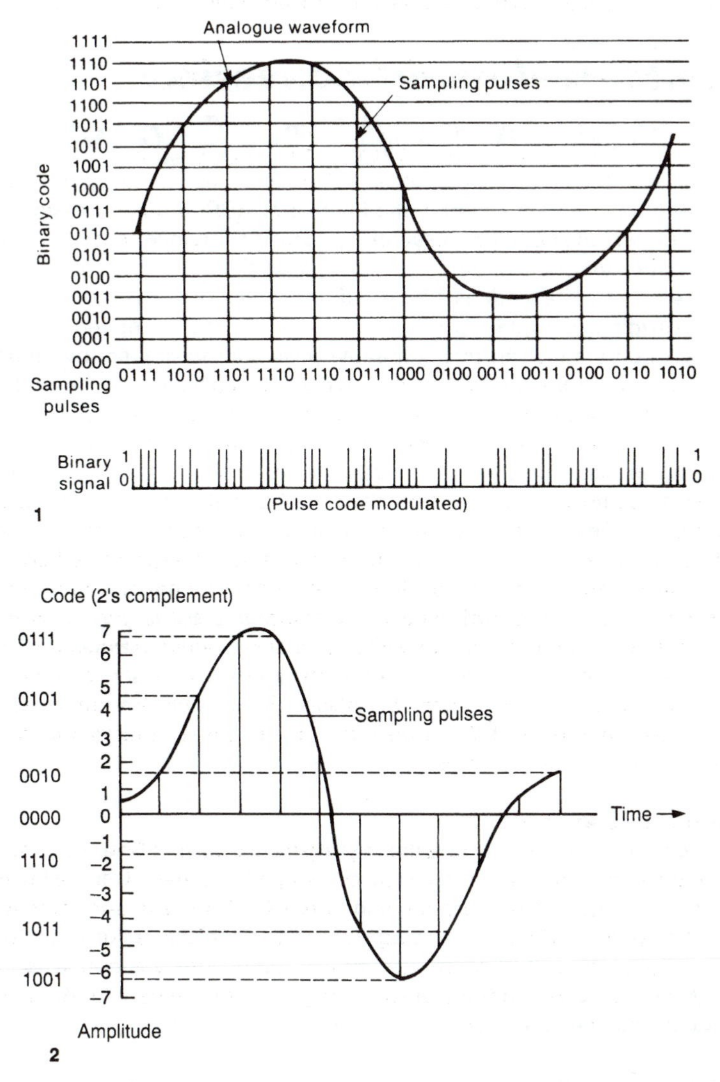

1. An example of sampling and coding of an audio signal. For simplicity only a 4-bit code is shown, which would allow only 16 levels of measurement (quantizing levels). This would result in a high degree of error and thereby a high level of quantizing noise. This takes the form of 'white noise' (i.e. noise equally distributed over the audio range) on large amplitude signals and 'granular distortion' (caused by intermodulation with harmonics of the sampling frequency) at low levels, where relatively few quantizing steps are in use and the noise is not masked by the audio. Some systems use non-linear quantizing so that there are more quantizing steps at low level where it is most critical.
2. The curve in (1) represents offset binary coding, where all the values are positive. Audio signals have waveforms with differing positive and negative values (above and below zero) so bipolar quantization is more appropriate. This curve shows a bipolar quantization system called *2's complement*, in which the negative values are indicated by complementing the positive codes and adding one.

Digital-to-analog conversion (DAC) is the inverse of ADC.

Digital-to-Analog Conversion, Oversampling and Error Correction

Digital-to-analog conversion is the inverse of ADC. In fact in many cases the DAC converter consists essentially of an ADC converter connected in a feedback loop.

In DAC the digital code is converted to a series of levels corresponding to the original samples, presented at the original sampling rate. The analog output is passed through a low-pass filter to prevent overlapping with harmonic images of the sampling frequency and to smooth the analog waveform.

Nyquist limit

The Nyquist theorem dictates that the sampling frequency must be at least twice the highest frequency it is intended to convert. In practice it must be substantially higher in order to allow for a practical filter to remove any out-of-band audio which would beat with the harmonics of the sampling frequency. This would produce an effect called *aliasing* which can be heard as spurious noises in the sound. If the sampling frequency were only slightly higher than twice the audio signal a very steep 'brick-wall' filter would be required to prevent aliasing. This could introduce serious phase distortion, but to increase the initial sampling rate would be expensive.

Oversampling

The above problem can be solved by increasing the effective sampling rate in the DAC to many times the original sampling frequency by a process of multiplication, using digital filters. For example, four times oversampling (as used in many compact disc reproducers) shifts the first harmonic of the sampling frequency to the spectral position where the fourth harmonic would have been, so that a comparatively gentle filter of 12 dB per octave would be adequate to prevent aliasing.

Error correction

Serious blemishes in digital audio reproduction, such as could be caused by drop-out on magnetic recording or dirt on CDs, can produce quite startling effects, but these can be identified and concealed or corrected by a process of *parity checking*. This consists of adding a parity bit to each binary number to make the number of 1s even (or odd). If an inversion has occurred the number of 1s will not be even (or odd). The parity check will detect this and will either repeat the previous sample or give a value between the previous and following value. This is known as *error concealment*. *Error correction* actually reconstructs the missing data by sending it out in blocks with parity checks interleaved, so that a burst of errors would be spread along the data stream.

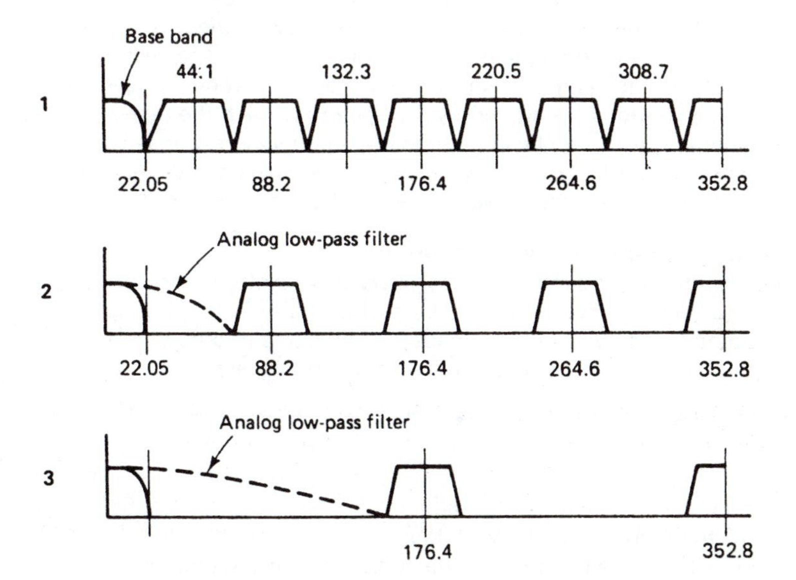

Digital oversampling filter. 1, the unfiltered response of a DA converter contains an infinite series of images of the original signal audio-band spectrum. 2, digital filtering with two times oversampling. 3, four times oversampling.

Oversampling suppresses the images immediately above the audio band and the remaining images can be filtered out with a relatively less steep filter.

A typical four-times oversampling filter could consist of a shift register of 24 delay elements, each delaying a 16-bit sample for one sample period during which it is multiplied four times, with a different coefficient used for each multiplication. The four sets of coefficients are applied to the samples in turn, producing four output values per sample. The 24 multiplication products are summed four times during each period and the product becomes the output from the filter, effectively increasing the sampling frequency by a factor of four. The output of the digital filter is converted to analog and the remaining band around 176.4 kHz is completely suppressed by an analog filter which need only have a 12 dB per octave slope. This can be achieved without phase distortion.

The principal difference between recording audio in analog and digital format concerns the different bandwidth requirements.

Recording Digital Audio

The main problem concerned with recording digital signals stems from the very large bandwidth required. To record an audio range of 20 to 20 000 Hz in digital form involves a bandwidth capability of between 1 and 2 MHz (derived by multiplying the sampling frequency by the number of bits in each sample and adding some extra to allow for error correction). This is well outside the range of normal audio tape recorders, where the high frequency response is limited by the speed at which the tape passes the recording/reproducing head and the low frequency response by the head gap size, although specialized tape recorders have been developed which can achieve the necessary packing density by special means (see p. 202).

Audio recording on video recorders

One method of obtaining the necessary bandwidth is to record digitized audio on a video recorder, thereby making use of the high tape-to-head speed afforded by the video heads, which rotate in a head drum at high speed, recording or reading the tracks diagonally across the tape.

Audio/video (PCM) adaptors

Before a video recorder can be used for recording digital audio, the signal has to be suitably processed. PCM adaptors are available which convert the audio into a pseudo-video signal, complete with synchronizing pulses to lock to the format of the VTR. Black level represents binary 0 and about 50% of peak white binary 1. As the audio signal is continuous whereas the video signal is interrupted by sync pulses, the gaps have to be covered by feeding the digital samples into a memory, to be read out at a rate which is higher than the sampling rate and then serialized in the correct order.

Editing

Video tape cannot be cut and spliced like audio tape. The proper relationship betweeen the sync pulses and the recording has to be maintained and anyway splicing helical scan tapes is not practicable because of the long slanting tracks. So video and digital tapes are edited by dubbing from one machine to another. This requires a sophisticated electronic synchronizing system which involves recording a digital time-code on the cue and address tracks of both machines. The machines can then be locked together, maintaining the proper sync pulse relationship throughout. The new sequence can then be butt-joined or cross-faded to the previously recorded one and the levels adjusted in the process. The digital recording suffers no degradation in the dubbing process.

Cue and address code

Using a standard VCR the audio signal can also be copied to the normal (longitudinal) audio track to help in identifying the sequences; the address code, on another longitudinal track, can provide a digital display of tape position.

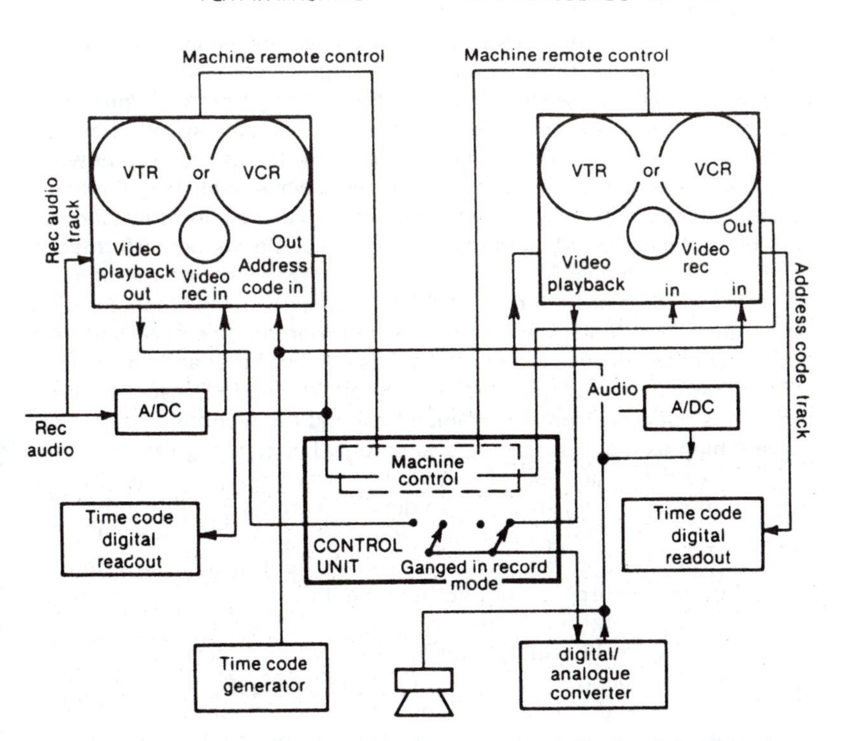

Schematic representation of two-machine time-code editing for digital sound. The control unit controls the mechanical functions of the playback machine and the record machine. A time code recorded on the address code track of each tape gives visual indication of their position. When both tape machines are locked to the synchronizer in the control unit they can be run in replay, record or spool in perfect sync. The synchronizer can also be used to locate a particular position on the tape.

Before editing, the machines automatically set to a position a few seconds ahead of the editing point. Editing is achieved by running the two machines together and playing in the selected sequence at exactly the required edit point on the master tape set by the time code. The edit can be checked by running through the transition in the 'rehearse' mode, when only the monitoring changes over at the edit point. If this is not exactly right, the edit point can be moved (by offsetting the electronic lock-up) and only when it is considered perfect is the device put into the 'record' mode. When the edit point is reached, the record machine goes into record and the edit is completed.

DASH recorders offer professional quality with flexibility.

Digital Audio Stationary Head (DASH) Recorders

Specialized digital audio recorders have been developed which offer digital quality and meet the requirements of professional recording. These include: multi-track recording, cut-splice editing, punch-in/punch-out editing with the ability to rehearse edits, variable-speed playback, independent control of tracks (to control which tracks record and which replay at the same time), off-tape monitoring while recording, the ability to handle various tape speeds, autolocation and the facility to synchronize with other machines. Most of these functions could not be performed by rotary-head machines.

DASH recorders are similar in appearance to analog tape recorders. The tape transport is fundamentally similar but more care is necessary over tension control because the digital tape is much thinner and smoother in order to provide good tape/head contact for extended HF response. Print-through is not a problem with digital tape and the recording density can be very high as the reproducing system only has to recognize two states of magnetism on the tape (S-N and N-S). The smoothness of the tape tends to trap air during spooling, so digital recorders do not spool as fast as analog ones. The capstan is controlled at constant speed when recording unrecorded tape, but on repro is controlled by a reference signal from a synchronizer or another tape; this is recorded on a control track, so that several machines can be run in synchronism. The control track also provides information for autolocation and specifies the type of format and sampling rate in use. Three sampling rates are available: 32 kHz, 44.1 kHz and 48 kHz. Error correction and timebase correction are used. The data code is cross-interleaved with delays so that dropouts and other tape errors are spread along the tape.

Punch-in/punch-out editing

The process of punching-in new material to replace a section of recording or assemble-editing is complicated by the data interleaving, because each moment in time is represented by a spread of data along the tape. So a crossfader is used, which deinterleaves the reply signal and reinterleaves it afterwards. The edit can be rehearsed with only the monitoring changing over before putting the machine into record.

Cut-and-splice editing

This is possible with DASH recordings. The tape must be cut at right angles. Splicing upsets the code sequence, but this occurs at different points in the odd and even sequence, so that by interpolation it is possible to obtain the end of the old recording and the beginning of the new one simultaneously, enabling the crossfader to make a rapid digital crossfade while both are available.

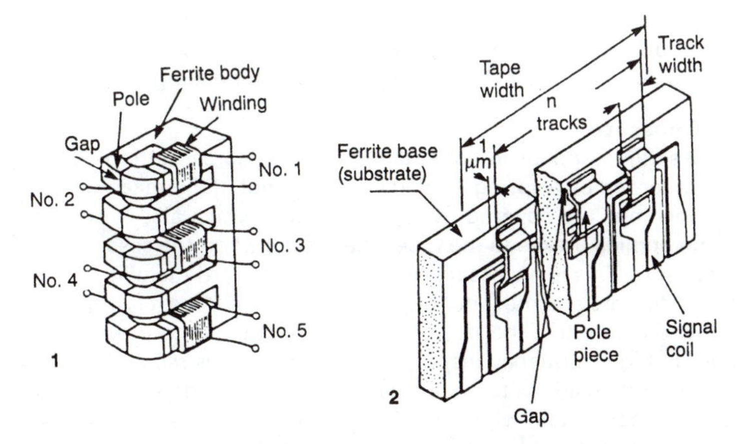

The problem of achieving sufficient bandwidth with a reasonable tape/head speed and multi-track capability is overcome by specially designed heads. Two types of heads are in use.

1. The ferrite head is used in the (single density) DASH 1 recorders (see below).
2. For DASH 2 (double-density) recorders, which have twice as many tracks, a special type of thin film has been developed, produced photographically by a method similar to the production of printed circuits.

There are several versions of the DASH format, including stereo and multi-track machines:

DASH-F is the fast version which runs at 30 in/s, providing 24 audio tracks on half-inch tape.

DASH-M is medium speed (15 in/s) with each audio channel spread over two tracks and is double-recorded, so that eight tracks provide two audio channels double-recorded on quarter-inch tape.

DASH-S is the slow version. Each audio channel is spread over four tracks with the tape running at 7.5 in/s, providing two audio tracks on quarter-inch tape.

Twin DASH means that the data for each audio channel are recorded twice, which makes the recording more able to cope with cut-splice editing.

DASH 2 is the double-density version, which uses thin film heads to achieve 48 digital tracks on half-inch tape or 16 tracks on quarter-inch tape. The tracks are so aligned that DASH 1 recorded tapes can be played on DASH 2 machines.

The machines have built-in time code generator/readers and the ability to lock to incoming standard time codes such as SMPTE/EBU.

The R-DAT Recorder

R-DAT stands for Rotary-head Digital Audio Tape. It records audio in digital form on a cassette measuring only 74 × 53 × 10 mm (2.9 × 2.1 × 0.4 in), which is smaller than a standard audio cassette. In standard format it offers up to two hours of stereo recording with very high quality, a signal/noise ratio better than 90 dB and excellent cueing facilities.

R-DAT cassette
The cassette is similar in construction to a video cassette, in which the tape is guarded by a spring-loaded flap when out of the machine. The hub drive openings are covered by a sliding panel which also locks the front flap and applies brakes to the tape wheel hubs when the cassette is not in the machine. This is necessary because the standard metallic tape is very thin and delicate and must not be touched by hand. Also it is important not to allow dirt to affect the head-to-tape contact so as to preserve the all-important high-frequency response. The cassette has recognition holes in four standard positions to enable cassette players to determine what type of tape is inserted and whether it is a pre-recorded tape. A recording lock-out plug is also provided to prevent accidental erasure.

Mechanism
The mechanism of the R-DAT recorder is similar to a helical-scan video recorder, with the tape wrapped around a rotating head drum, the tracks being recorded diagonally across the tape. The standard head drum is 30 mm diameter and rotates at 2000 rev/min (1000 rev/min in long-play mode). Smaller drums are used for portable machines with suitably altered speeds. The tape has only 90° of wrap around the head drum (video recorders have at least 180°). This makes for easy lace-up and reduces the risk of tape damage. It also makes for lower tape tension, with consequent reduction in head wear. The tape/head pressure is only 0.1 gram and the tape remains in contact with the heads during spooling, allowing for very rapid cue location. The read/write heads are set at 180° in the drum, but with only 90° of tape wrap they only make contact for 50% of the time, so the digitized audio signal has to be time-compressed by feeding blocks of samples into a memory at the sampling rate and then reading them out at a much faster rate so that the writing is done in less than real time.

Format
The standard format uses 13 μm thick tape with a speed of 8.15 mm/s which offers up to 120 minutes of recorded time with a sampling rate of 48 kHz. Using 7 μm tape can increase the playing time to 180 minutes. Domestic machines can record and replay at 48 kHz sampling rate, but pre-recorded tapes are played back at 44.1 kHz with a tape speed of 12.225 mm/s, allowing a maximum playing time of 80 minutes, quite adequate for most purchased material. Professional machines can also record at 44.1 kHz (see p. 206).

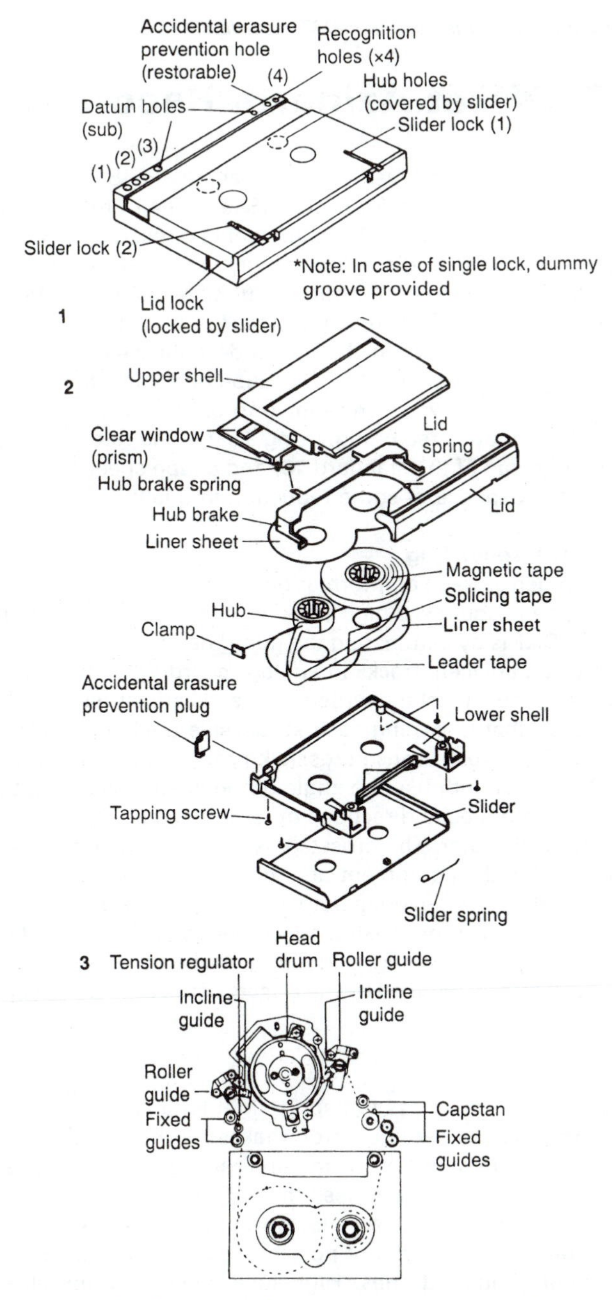

1. Bottom view of R-DAT cassette. The recognition holes identify the type of recording on the tape. (Top view: p. 145.)
2. Exploded view of R-DAT cassette. The clear window (prism) is built into the cassette for the light-actuated end-of-tape sensor (the tape has a clear leader section).
3. R-DAT transport mechanism.

R-DAT Offset Azimuth Recording

The standard for consumer-use R-DAT machines is 48 kHz sampling for recording and 48 or 44.1 kHz for replay. (See p. 204.) Professional R-DAT machines can also record at 44.1 kHz, to enable direct digital mastering of R-DAT tapes and compact discs. This ability is deliberately not provided on domestic machines to inhibit direct dubbing (copying). As a further discouragement each R-DAT machine records its serial number, along with positional information, in the sub-code data which is recorded at each end of the diagonal tracks. The sub-code can also be used for recording time-code for synchronization purposes. It is possible to provide four hours of continuous recording, using a sampling rate of 32 kHz, 12-bit quantization (instead of the standard 16) and a tape speed of 4.075 mm/s, but this is at the expense of some reduction in quality.

Offset azimuth recording

To obtain so much storage of information in so small a cassette with such a low tape speed requires a high packing density. One way that this is achieved in R-DAT is by eliminating the guard bands (the spaces normally allowed between adjacent tracks in analog recording), which would be a waste of tape space. In fact the write/read heads are larger than the width of the tracks, so that in writing each successive track they overwrite the previous one. This would cause crosstalk problems in analog recording, but in R-DAT the azimuth (i.e. the angle of the heads relative to the track) is offset by 20° in opposite directions on each head so that the tracks are laid at 40° to each other. The crosstalk is therefore minimal and as the recording is digital the equipment only has to discriminate between two states of magnetism. The overlap between the tracks results in their being butted up together with no wasted space between. The same heads are used for recording and reading, and although they overlap the tracks the crosstalk is sufficiently reduced by the offsets of the azimuths to overcome the system noise.

R-DAT operation

The combination of low-mass transport, light tape and tension, together with very sophisticated servo control, makes the R-DAT recorder particularly suited to rapid selection of material and editing. It is possible to rewind the whole of the tape in less than 15 seconds and park it within 1/48 000 seconds of the required cue, where it can be put into record or replay immediately. It is therefore possible to make very quick jump-cuts between pre-programmed cues. High-speed search is possible at 200 times normal. Editing between machines, however, involves a similar arrangement to that for editing PCM-adapted video machines, i.e. it can only take place at the right point in the track sequence and the servos have to be synchronized and locked together at the change-over point (see p. 201).

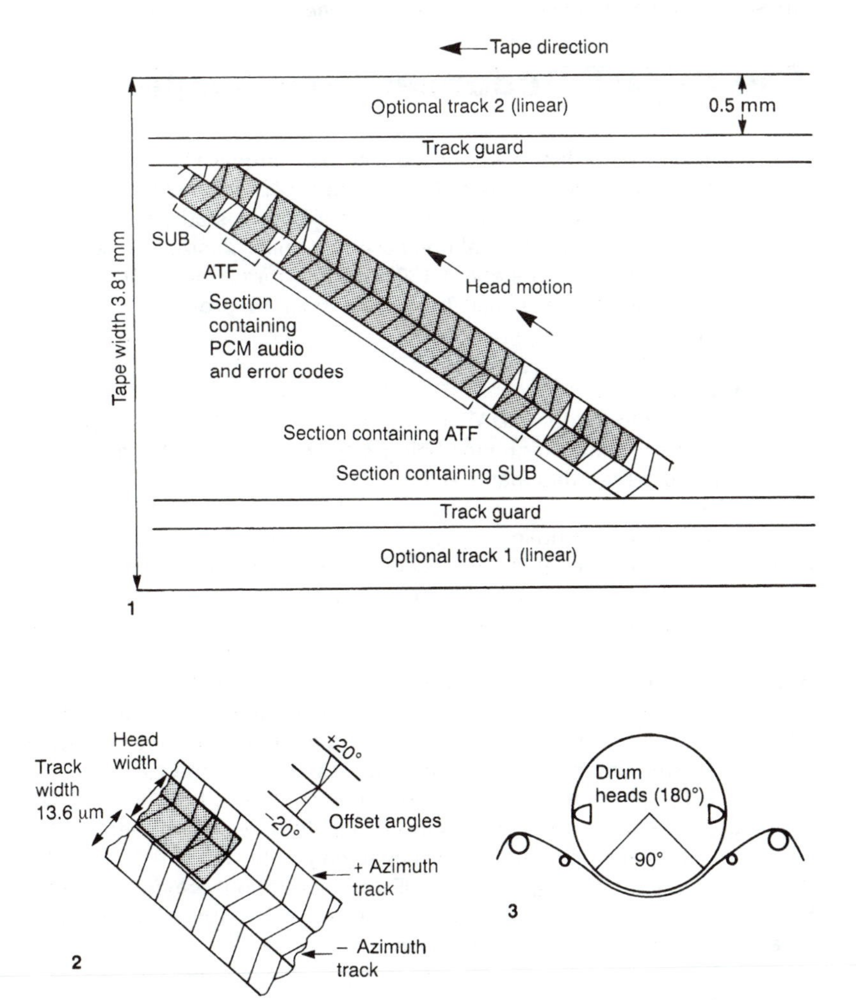

1. The R-DAT diagonal tracks consist of a middle section containing the digital audio and error codes, at each end of which is a section containing blocks with signals for automatic track-finding (for high-speed search) and track-following and a sub-code section containing system data. This could include information about the contents of the recording, starting points and durations of each section etc. as well as the serial number of the recorder. Space is also available to include graphic information which could be displayed on a screen. The Automatic Track Following signal is of low enough frequency to be little affected by the offset azimuths, so that the crosstalk can be picked up and compared on both sides of the track and the tracking adjusted accordingly. Also provided are two linear tracks, one at each edge of the tape, but due to the slow tape speed they are of limited use and serve mainly to protect the diagonal tracks from edge damage.
2. Enlarged view of tracks, showing th rec/rep head 1.5 times large than the tracks, so that there are no spaces (guard bands) between the tracks.
3. Head-drum formation, showing 90° wrap and heads set at 180°.

The S-DAT recorder has many of the features of the R-DAT machine but with a static head stack in place of the head drum.

The S-DAT Recorder (prototype)

S-DAT cassette recorders are being developed largely for the domestic consumer market. One type is a digital tape recorder working on rather similar principles to the R-DAT recorder, but with a static, instead of a rotating, head. It uses a self-sealing cassette, similar in size to a compact audio cassette, which holds about 130 m of 3.78 mm wide by 10 μm tape and is double-sided, i.e. it can be turned over to provide double the playing time.

Operating modes

The operating modes for S-DAT are rather similar to R-DAT (see p. 204), the standard sampling frequency being 48 kHz, with 44.1 kHz being used for replay of pre-recorded material. In one version the playing time for the standard option (two channels) is 90 minutes (using 10 μm tape), 135 minutes with 32 kHz sampling and 180 minutes at half speed. Pre-recorded tapes have a maximum of 38 minutes on each side of the cassette. A four-channel option is also available, with a sampling frequency of 32 kHz and a playing time of 90 minutes.

Head arrangement

S-DAT recorders have a static head stack instead of the rotating drum of the R-DAT machines. With a linear tape speed of only 47.6 mm/s (23.8 mm/s in the double-play mode) special measures are required to obtain the necessary digital data-packing density. This is achieved by dividing the digital codes between many tracks in parallel. In one version there are 20 tracks on each sector of the tape, occupying only half the tape width (1.905 mm). Each track is only about 65 μm wide (about the diameter of a human hair) so special head stacks are used, which are of the thin-film type, made by a photographic process.

Magnetoresistive heads

With tracks only 65 μm wide and a linear speed of only 47.6 mm/s it would not be possible to obtain sufficient output by the normal magnetic reproducing process (where the varying flux on the tape generates a current in the head coils as a function of the speed of the tape as it passes the head). So special magnetoresistive heads are used, in which the resistance is governed by the strength of the flux on the tape, but is not dependent upon its rate of change. Instead of the tape generating the signal, it modulates it. The signal level is so small that miniature integrated-circuit amplifiers have to be mounted in the heads. In the recording mode the head adopts the more usual inductive role to convert the incoming signal into magnetic variations on the tape.

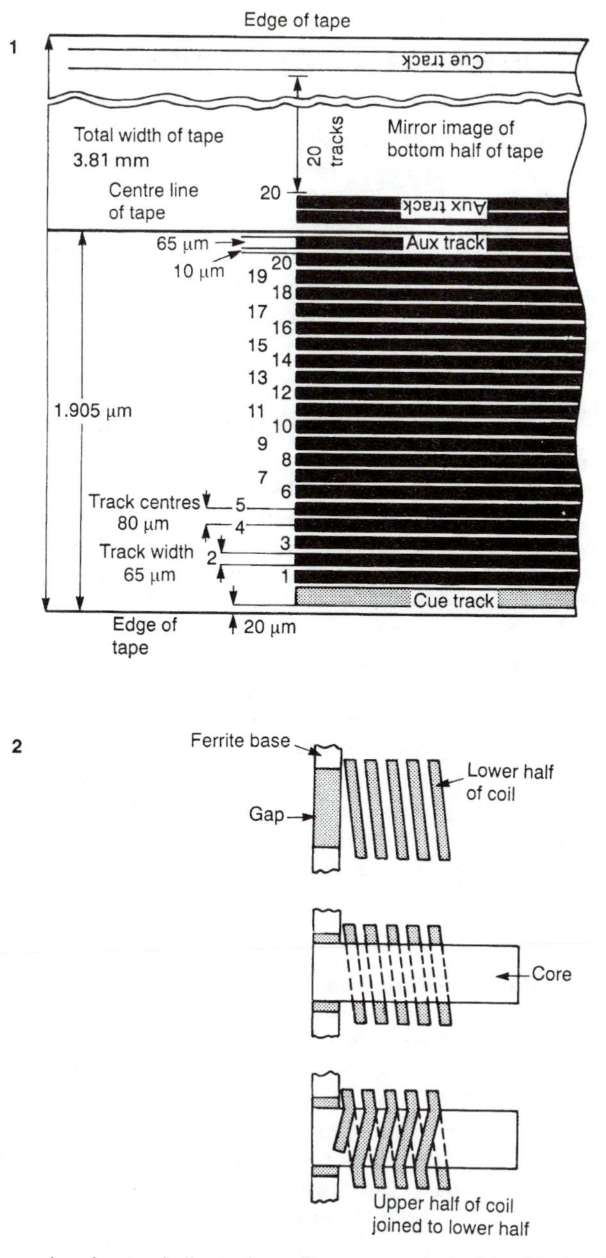

1. S-DAT tape, showing track dimensions. The cassette is double-sided, only half the tape width being used at a time. Turning the tape over doubles the playing time.
2. Construction of a thin-film head, which is made by a photographic process. The conductors are deposited as flat strips on the film with the core sandwiched in between. The strips are joined at the edges to form a coil.

The Philips DCC system provides the ability to play back digital or standard analog audio cassettes on the same machine.

The Philips DCC Recorder

The Philips DCC recorder provides for digital recording and playback as well as playback-only of analog recordings using standard-sized compact audio cassettes. It works on the S-DAT principle (see p. 208), i.e. instead of using a rotating head, as in R-DAT, in this case the necessary digital packing density is achieved by dividing the digitized signal between eight parallel tracks. A ninth track is also provided for control and it is possible to accommodate up to 400 characters of text per second for visual information.

The cassette

The DCC cassettes are similar in size to standard audio cassettes, but incorporating a sliding metal cover to protect the tape and guides, which slides away in the manner of video cassettes when inserted in the machine. They do not require a separate case, the detail labels being stuck on the top side. Pre-recorded DCC cassettes have additional packaging to accommodate sleeve notes.

The tape

The DCC tape is similar in formulation to video cassette tape. Like audio casette tape it is 3.78 mm wide and plays at the standard 4.7 cm/s (1⅞ in/s). The nine parallel tracks (including the one used for control information), which are each only 185 microns wide, occupy only half of the tape width, so that the tape can be played in both directions, allowing for 45 minutes playing time each way (possibly increased to 2 × 60 minutes using thinner tape). However, instead of having to turn the cassette over to play both sides the head turns over as the tape reverses. This is achieved very rapidly, enabling any point in the programme to be available for rapid selection.

The record/playback function

DCC machines can record and replay in the digital mode, but only replay analog recordings. There is no erase head, previous digital recording being overwritten as new recordings are made. The system will work with sampling rates of 48, 44.1 or 32 kHz and this is selected automatically to match the input digital signal. Error detection and correction codes are inserted. The frequency response for the two higher sampling rates is equivalent to a compact disc, with a dynamic range of 108 dB and total harmonic distortion, over the frequency range 5–20 000 Hz, of –92 dB. Pre-recorded digital recordings are to include the 'Serial Copy Management System' to inhibit digital copying.

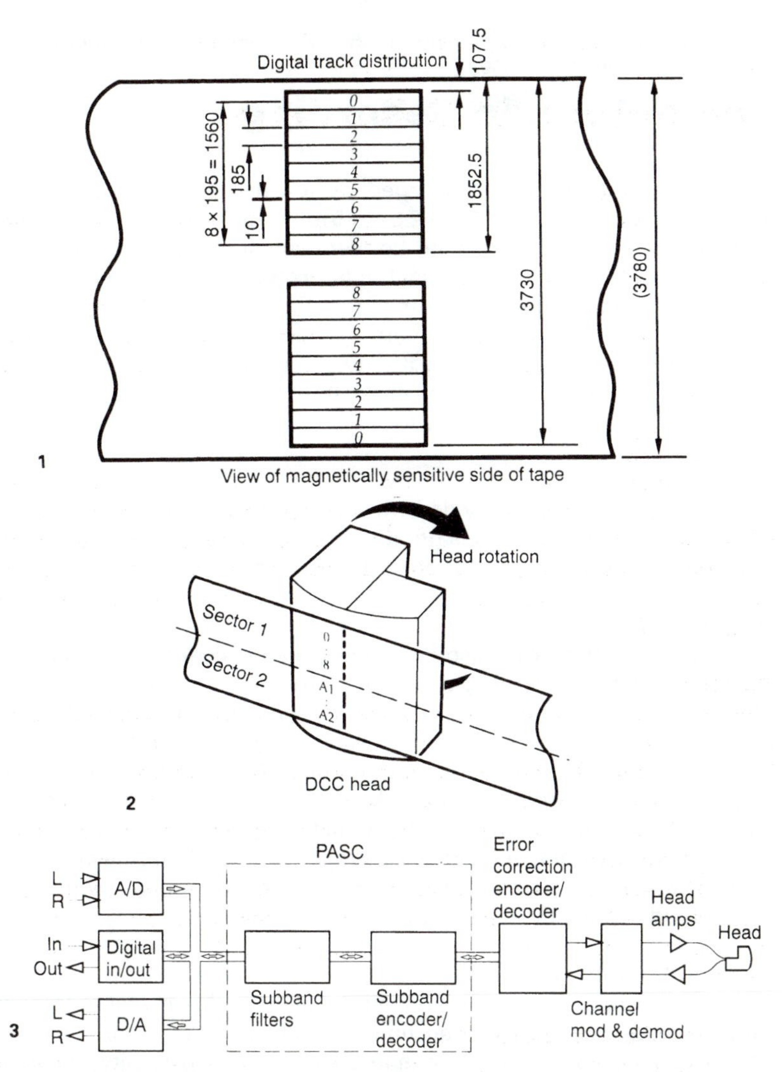

1. Lateral dimensions of the tape, as viewed from the sensitive side. The tracks as recorded are 185 microns wide, but the playback head gap is only 70 μ wide to allow for tape weave. The combination of very narrow track and relatively slow speed would result in too small an output if magnetic induction were employed, so magnetoresistive heads are used (see p. 208).
2. The DCC combined digital/analog tape head.
3. An outline of the DCC coding system. To achieve the necessary data-packing density a system called PASC (Precision Adaptive Sub-band Coding) is used. The audio frequency band is divided into 32 sub-bands, each being continuously assessed and allocated only the minimum requirement of bits. Signals below the threshold of hearing (which is continually adjusted to take account of the 'masking' effect of loud sounds on quieter ones) are ignored and spare bits allocated where increased data accuracy is needed. This results in an average bit-rate of only 4 bits per sample, rising to 15 bits where needed, and is equivalent to 18-bit coding overall. The 32 sub-bands are matrixed into an eight-channel bit-stream for the eight-track head.

The Video 8 format provides for high-quality PCM audio with video recording or long-play audio-only recording.

Video 8 Audio Recording

The Video 8 format was developed as a video recording system. It employs a small cassette (95 × 62.5 × 15 mm), much smaller than a VHS cassette and particularly suited to very small video recorders, e.g. camcorders (integrated camera and video recorders). It is a helical-scan recorder with a rotating head drum supporting two heads, rather like the R-DAT recorder (see p. 204) but with the tape making a half-cylindrical wrap around the drum. There are also two linear audio tracks, but as they are narrow and the tape speed is only 20.05 mm/s (10.06 mm/s in the long-play mode) they are not suitable for high-quality recording.

FM carrier

An FM carrier system is provided which supports a single audio channel in the video waveform. This is good quality but monophonic and cannot be divorced from the video for editing purposes.

PCM recording

The Video 8 system can also offer PCM audio recording in stereo to accompany video by writing the audio coded signal in a section of the tape where there is no video. This is achieved by extending the half-cylindrical wrap by a further 31°, in which space the audio signal (which is time-compressed in a similar manner to the R-DAT recorder) is recorded. As this extends the length of the tracks a 1.25 mm wider tape is used. The drum contains two read/write heads so that while one head is reading the last 31° of a video field the other will be reading the audio at the beginning of the next. As the video and audio signals are coded separately they can be recorded separately, possibly at different times.

8 mm audio-only recording

The Video 8 format can be used as a high-quality audio-only recorder, with the possibility of a long playing time. The ability to time-compress the audio signal into 31° of head rotation opens up the possibility to extend the playing time by recording six audio segments in one track. Only one segment in each track is recorded as the tape passes, but when it reaches the end of its travel it reverses and records the next segment in the opposite direction until all segments are filled.

The Video 8 format has no control track. Track following is achieved by signals contained in the recording. It is therefore not possible to play back more than one segment at a time as there is no fixed relationship between them, so the system cannot be used as a multi-track recorder.

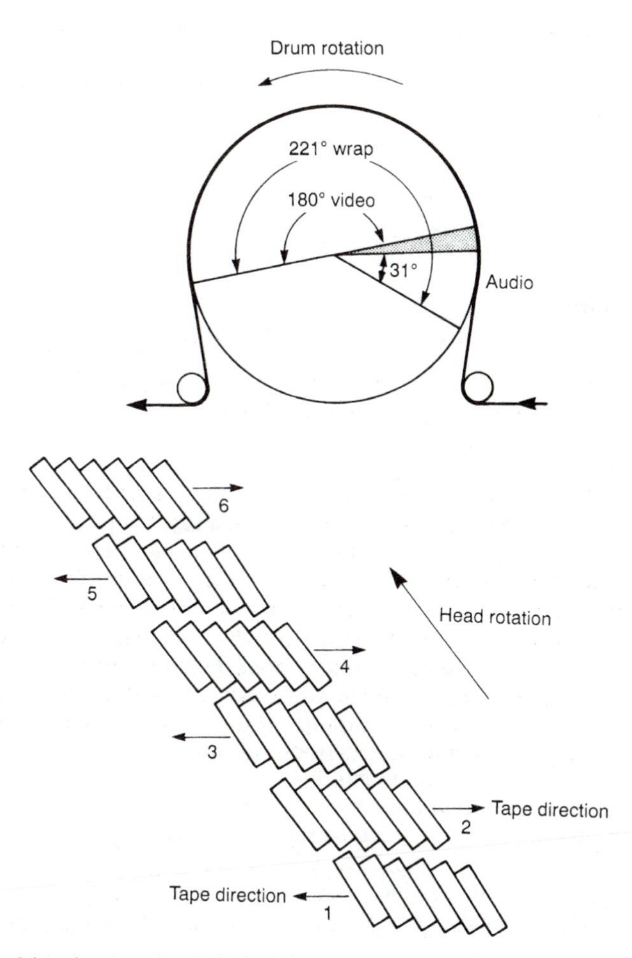

1. Video 8 head wrap with PCM audio. Video occupies 180° of drum rotation, the time-compressed digital audio a further 31°. A 1.25 mm wider tape is used for the PCM audio version to accommodate the extra length of the tracks.
2. 8 mm audio-only recording. Six audio-only recordings are made in alternate directions, the tape reversing between each. As the head-drum continues to rotate in the same direction the tracks do not line up with each other and are recorded at a slightly different speed, so only one can be used at a time. Controls for tracking and tape direction are contained in the coded signal.

Compact Discs

The compact disc is a digital audio system, normally used for reproducing pre-recorded music. The standard compact disc is 12 cm in diameter and is played on one side only by a non-contact system, involving a laser beam focused on the disc from underneath. The standard disc provides a maximum playing time of 74 minutes. There are also 8 cm (singles) discs with a maximum playing time of 20 minutes and three types of discs that provide video as well as sound, some of which are played on both sides (see LaserVision in Glossary). Some players accept multiple discs, loaded in a magazine or 'carousel'.

High quality
The compact disc offers almost perfect reproduction of the recorded material, with a frequency response within ±0.5 dB from 20 Hz to 20 kHz and a dynamic range, signal/noise ratio and separation between channels all better than 90 dB. As it is a digital system, controlled by a quartz crystal, wow and flutter are virtually non-existent and as there is no physical contact between the disc and the reproducing system there is no deterioration due to wear. Moreover the discs are quite robust and, provided that efficient error correction is applied, can be handled and allowed to accumulate a small amount of dust, or even scratches, without obvious effect.

Manufacture
The compact disc master is usually made from a magnetic disk tape recording, the actual technology involved being identified on the label by a three letter code in earlier recordings. It is no longer always applied.
DDD denotes that digital tape recording has been used for the recording session, mixing, editing and mastering.
ADD denotes that an analog tape recorder has been used for the recording session, with a digital tape recorder used for subsequent mixing and/or editing and for mastering to CD.
AAD denotes that an analog tape recorder has been used for the recording session, mixing and editing, with a digital recorder used for the final transcription to CD.

Recording
The master recording, from which CDs are made, would normally supply audio in digital form with a sampling rate of 44.1 kHz and 16-bit resolution for each of the stereo channels. EFM (eight-to-fourteen modulation: see Glossary) is used in order to reduce the required bandwidth and the signal's DC content, while providing an element of 'self-clocking'. The compact disc master is made photographically, using a laser to produce pits in a photo-resist surface coated on a glass disc. In the production process the pits become translated into bumps on the underside of the consumer discs, which have an aluminized reflective surface. This is then coated with a 1.2 mm layer of plastic and the label stuck on the other side.

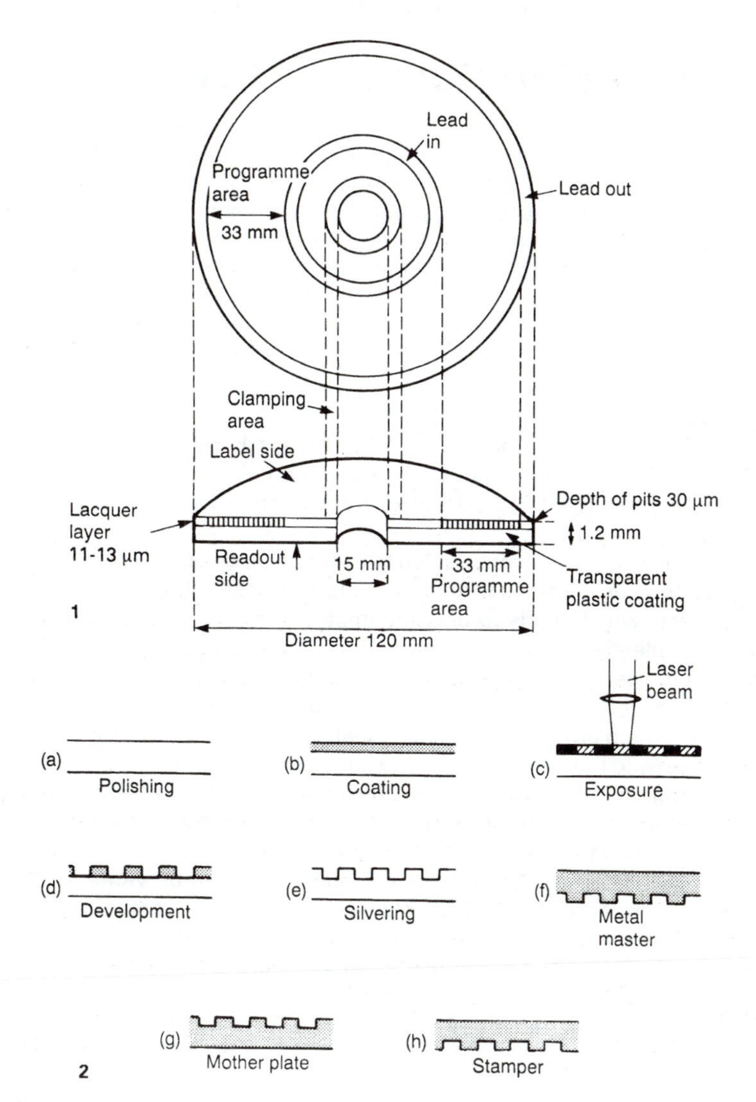

1. General dimensions of a standard compact disc.
2. Outline of the processes involved in the manufacture of compact discs. (*a*) A glass plate is polished for smoothness. (*b*) A photo-resist coating is applied to the polished surface and cured in an oven. (*c*) The coating is exposed to a laser beam, the intensity of which is modulated by the digital audio signal (usually from a digital tape recorder). (*d*) The photo-resist surface is developed to produce the pits. (*e*) The surface is coated with silver by evaporation to protect the pits and make it conductive. (*f*) The surface is then plated with nickel to produce a metal master. (*g*) From this master mother plates are made. (*h*) From the mother plates stampers are made which are used to produce the consumer discs in large numbers by an injection or compression moulding process. Finally the discs are coated with a protective plastic coating and the label added.

CDs are played by a laser beam focused on their reflective surface through a protective coating.

Reproducing Compact Discs

The manufacture of compact discs (page 214) produces pits in the surface of the master disc which are reproduced as bumps on the underside of the consumer discs. The length of the bumps and the spaces between them form the digital information on the disc. They are arranged as a clockwise spiral track which starts at the inside of the disc and works outwards. The track pitch is only 1.6 μm wide (about 1⁄40 of the diameter of a human hair and about 1⁄60 of the average groove spacing of an LP record). The spiral track of a standard CD is about 5.7 km long.

Readout

The digital coded information from the bumps is read by a laser beam which is focused very accurately on the reflective surface between the bumps through the 1.2 mm layer of transparent plastic which coats the underside of the disc.

The optical system is constructed to make the laser beam come to a focal point sharply as it passes through the plastic coating. It is 0.8 mm wide at the surface, which tends to prevent small particles of dust and scratches from completely interrupting the beam. Where the laser beam hits the spaces between the bumps it is reflected back along the original optical path, where it is intercepted by a beam-splitting prism and directed to a photo-sensor to provide the output signal. Where the beam hits a bump it is defocused, and the dimensions of the bump in relation to the wavelength of the laser light are such that the reflected beam is caused to scatter and very little is returned to the photo-sensor. As it is a digital signal the output only has to represent the transitions between 'on' and 'off'.

Error correction (p. 198) is applied, making CDs impervious to minor blemishes on the disc. Most CD players incorporate 'oversampling' (see p. 198) to improve the digital-to-analog conversion. Four-times over-sampling is typical but some use a higher figure and some a system of Pulse Density Modulation or 'bitstream' conversion which converts the 16-bit digital signal into a high-speed 1-bit data stream (see Glossary).

Subcodes

In addition to the programme recording there is provision in the digital signal for eight separate subcodes (nominated P, Q, R, S, T, U, V and W). The P subcode is a music track flag indicating the start of each section and the lead-out at the end. It can be used for search systems. The Q code contains such information as: table of contents, track number, starting time and running time for each track, etc. It is also used to signal the number of tracks used (four tracks are possible) and to switch in the appropriate degree of pre-emphasis. The other subcodes can be used for video (see Glossary – LaserVision).

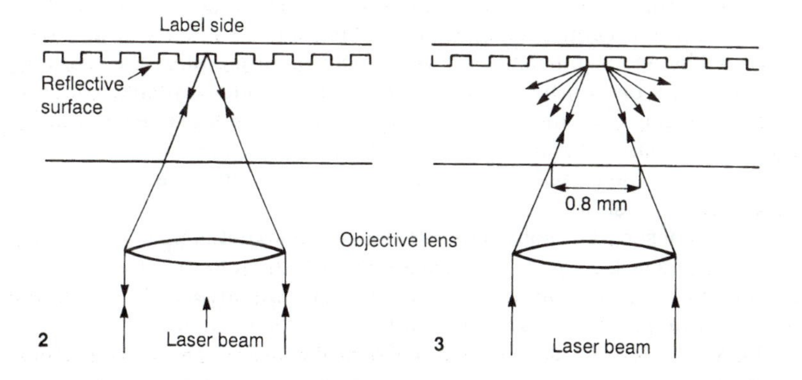

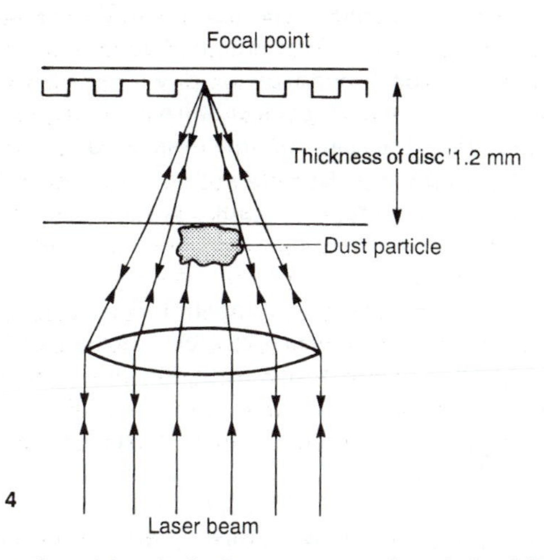

1. Dimensions and spacing of the pits (or bumps as seen from below). The spiral of pits starts at the inside of the disc and works outwards. The rotational speed of the disc is varied continuously to maintain constant velocity at the reading point.
2. The laser light beam is focused with great accuracy on the spaces between the bumps. As this is reflective the beam is reflected back along its original path.
3. When the laser beam hits a bump it is defocused and scattered, so that much less is reflected back along the original path.
4. The laser beam is brought to a focal point sharply, and as it passes through the outer surface of the disc it is 0.8 mm wide. This means that small blemishes or dust particles of lesser dimension only form a partial shadow and do not obscure the beam completely.

CD replay involves very sophisticated optical tracking and focus mechanisms.

Compact Disc Tracking

Focus

The CD replay mechanism involves extreme accuracy in the focusing and tracking of the laser beam, even to the extent of being able to cope with some warp or eccentricity in the disc. As there is no stylus contact this is achieved by optical means. In a typical focus control system the returning light beam is passed to the sensor through a cylindrical lens. The lens, being cylindrical, magnifies in one plane only and has the effect of turning the round beam into an ellipse on either side of the focus point. The orientation of the ellipse changes according to whether the focus error is too long or too short.

The signal sensor is split into four segments so that opposite quadrants receive more or less of the elliptical beam according to the direction of its orientation. The outputs of the sensors are differentiated to produce a signal to control the focus mechanism. Perfect focus is indicated when the outputs of all four sensors are equal. Control of focus has to be very rapid to accommodate possible warp on the disc. A mechanism not unlike a moving-coil loudspeaker can be used, with the objective lens taking the place of the centre cone of the speech coil, and the beam shining through an aperture in the centre of the magnet pole piece, coincident with the lens. The lens is thus moved up or down by the current through the moving coil. As the range of focus capture provided by the cylindrical lens method is restricted it can be necessary to employ a focus-search mechanism to ramp the focus servo up and down over its whole range of travel to find the focus crossover point when the disc is first inserted.

Tracking

A typical method of obtaining accurate tracking uses a diffraction grating (a metal plate with thin parallel slits spaced a few wavelengths apart) in the laser beam to split it into three, with the two additional beams being focused on the disc one on each side of the centre line of the track. The reflected light from the two side beams is returned by the beam-splitting prism to the photosensor and the outputs of the two side sensors are compared; the difference signal is used to direct the tracking mechanism.

The focus and tracking mechanisms are combined into a two-axis device which consists essentially of two coils set at right angles in magnetic fields. Control currents in the coils move the objective lens vertically for focus and horizontally for tracking.

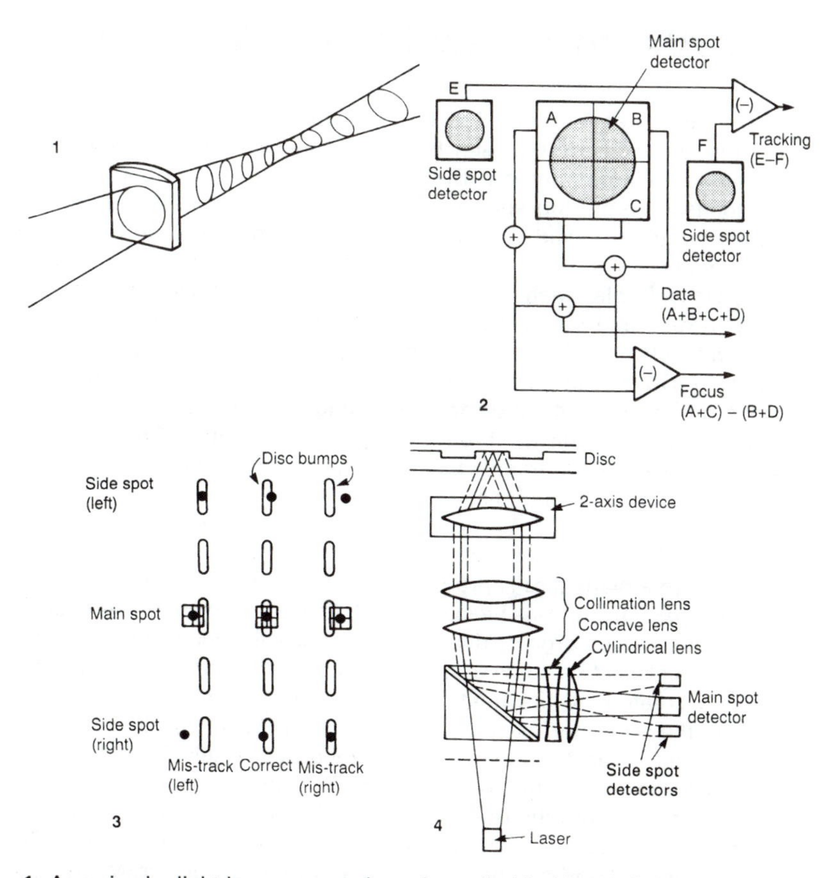

1. As a circular light beam passes through a cylindrical lens it forms an ellipse, the orientation of which varies with distance from the lens. The beam, reflected from the disc and diverted by a beam-splitting prism, passes through a cylindrical lens to the photosensor.
2. The photosensor is composed of four quadrants with the outputs of opposite segments added together and then subtracted one from the other to provide the focus control signal. When the light beam is in perfect focus it is circular so that the output from all four segments is equal and the focus control signal is zero. Short or long focus error produces a difference signal of the appropriate polarity which drives the focus mechanism. The sum of the outputs of all four sensors provides the data readout.
3. Tracking is controlled by two side spots. They are projected to each side of the centre of the track and their reflected beams are picked up by two side spot detectors arranged on either side of the main photosensor. Subtracting their outputs provides the tracking error signal.
4. General arrangement of the optical system. The 2-axis device moves the objective lens vertically for focus and horizontally for tracking. In the 'flat-type' optical pick-up a mirror prism is incorporated to turn the beam through 90° to provide a low profile.

Some types of optical discs can be recorded by the consumer.

Recordable Optical (WORM) Discs

Compact discs are an excellent recording medium where multiple copies are required as, once the master blank has been produced, they can be stamped out in large numbers. But the production of CDs is complicated and expensive, requiring specialized equipment and scrupulous manufacturing conditions, and is commerically viable only for large volume production.

Write Once, Replay Many times (WORM) discs, which can be recorded directly in real time by the consumer, provide the ability for immediate replay with relatively low cost for small numbers, making them a good medium for CD mastering (eliminating the use of tape), promotional and demonstration purposes and for effects, jingles and music 'stings' and similar cue provision in the broadcasting and film fields. They are also particularly suitable for archive purposes, having a shelf life in excess of ten years in normal conditions.

CD-R

In this system a blank disc is provided consisting of a polycarbonate disc in which a plain groove has been stamped. The grooved surface is coated with a light-sensitive organic dye followed by a reflective coating of gold and finally a protective coating of UV lacquer. The pits that comprise the signal are created by the heat from a powerful laser beam (6–8 mW instead of the more normal 1–2 mW used for playback) which is focused on the organic layer. The intense heat formed by the spot from the laser causes the organic material to decompose, creating a minute gas bubble which forces itself into the polycarbonate substrate below the surface of the groove. The resultant recording, composed of pits and bumps, is perfectly standard and can be played on a normal CD player. Moreover, unlike the normal CD, which is produced by a lengthy manufacturing process, the recordable CD is available for use immediately after recording.

Metal discs

Another type of WORM disc consists of a thin metal disc in which the recording is made by a laser beam which is so powerful that it can melt the metal, creating holes in the disc where it is turned on by the digital signal. The recording is read by a low-power laser, the beam of which is reflected back from the disc except where it encounters a hole which it passes through.

Recordings made by these methods are permanent and the discs cannot be re-recorded. Recording from an analog source would require an A/D converter as well as an EFM (eight-to-fourteen modulation) encoder, as in the case of normal CD mastering, but copying from another CD (where the signal has already been through these processes) could be direct.

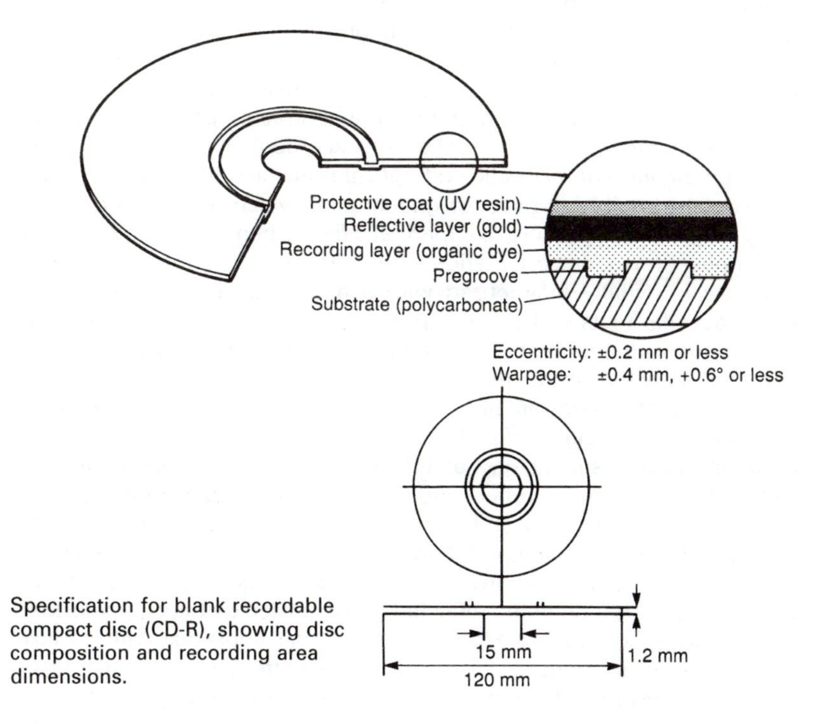

Specification for blank recordable compact disc (CD-R), showing disc composition and recording area dimensions.

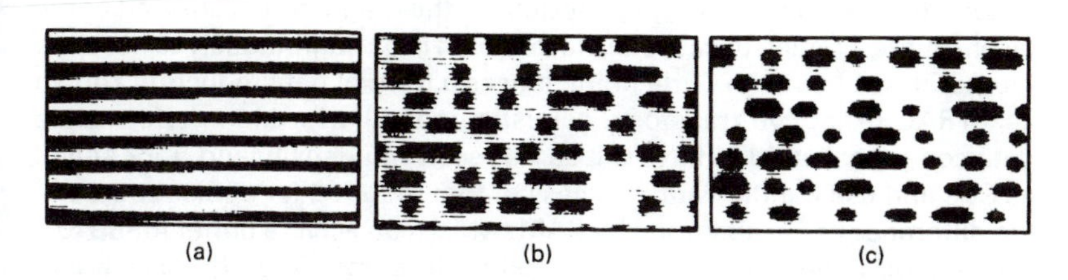

Microscopic view of CD-R, (*a*) before and (*b*) after recording, with (*c*) standard recorded CD for comparison.

The DRAW disc is available for playback immediately after recording and is reusable.

Recordable, Reusable (DRAW) Discs

The recordable compact discs described on page 220 have the possible disadvantage of only being recordable once and not reusable.

A form of disc which is both recordable and reusable is the DRAW disc. Unlike the CD it is available for replay immediately after recording (DRAW stands for Direct Read After Write), although erasure probably requires a separate operation and it cannot be over-recorded. The recording is made by a combination of magnetic and optical principles. The disc is made of a material (possibly terbium or gadolinium) which has the property of becoming very susceptible to a magnetic field when heated and retaining the imposed magnetic orientation when returned to normal temperature. The disc is initially magnetized with a magnetic flux in one direction (perpendicular to its surface). It is passed over an electromagnet which is continuously attempting to apply a flux in the opposite direction, but the strength of the field is insufficient to overcome the coercivity of the metal, so the magnetic state remains unchanged. On recording, a laser beam, modulated by the digital signal, is focused on the disc so that the pulses momentarily heat small points on the disc to beyond its Curie temperature (when its coercivity becomes zero) and it assimilates the reversed flux from the coil. When the heat is removed this condition is 'frozen' so that the disc retains a flux pattern that reverses in relation to the digital pulses. Readout is obtained by shining a beam of light on the disc and sensing the change in polarization of the reflected beam caused by the magnetic field. To erase the recording it is necessary to scan the whole disc with the laser on high power to produce the zero coercivity effect, while the field coils apply a flux reversed from the direction used for recording.

Magneto-optical MiniDisc

The Recordable MiniDisc, used for mastering, or as a digital alternative to NAB Cart machines, is similar in operation to the standard recordable CD except that the recording signal modulates the magnetic coil instead of the laser. This is simpler, because the laser beam is on continuously to keep the temperature above the Curie point. It uses a form of data reduction known as ATRAC (Adaptive transform acoustic coding) with high-performance data compression. MDs have the same recording density and track speed as CDs and can handle data in excess of that required for audio recorders, the difference being absorbed by RAM (which can act as a buffer for up to 3 seconds of audio). The RAM is topped up in bursts from the disc. It can cover any short periods of loss from the disc. The Recordable MiniDisc can have the same playing time as a pre-recorded disc.

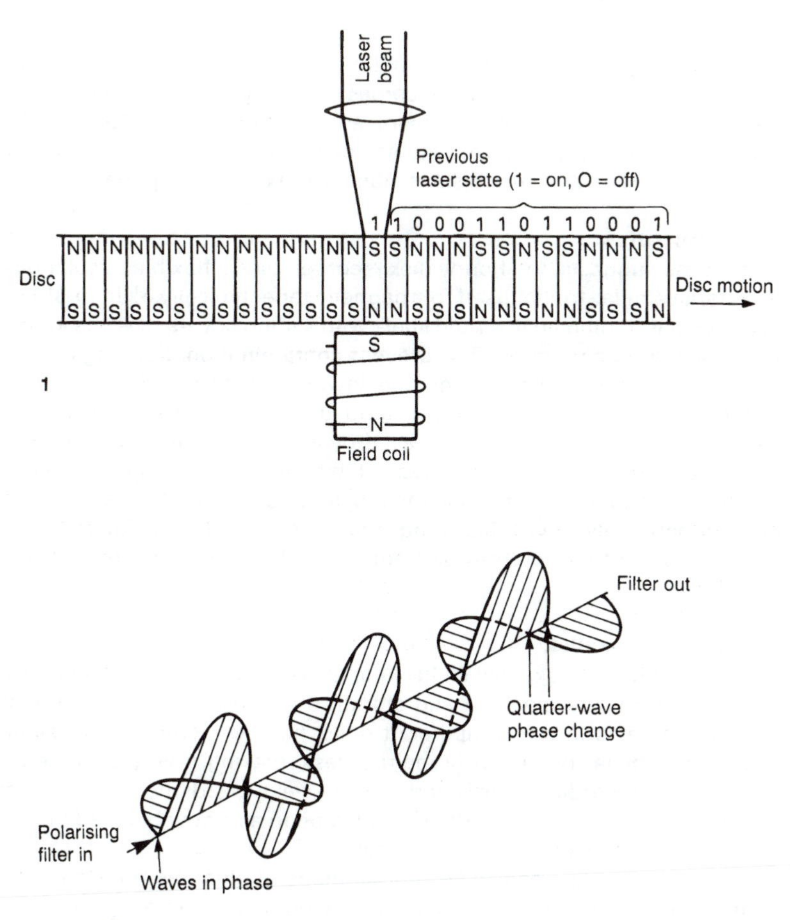

1. In its 'erased' state the thermomagneto-optic disc is magnetized with a single direction of flux all over. It is continually subject to flux in the reverse direction from the field coil, but this is not strong enough to influence its magnetic state until the disc is heated by the laser beam.
2. The disc is replayed by focusing a beam of polarized light on to the track and sensing the change in polarization in the reflection due to the magnetic field. A beam of light can be considered as an electromagnetic wave with two components acting at right angles to each other. When passed through a polarizing filter the two components are propagated with different speeds and the plane of polarization is rotated. The degree of polarization can be 'read' by passing the reflected beam through another polarizing filter.

Magnetic Disks

Tape recording can be an excellent method of storing audio, but it can also be cumbersome and slow when it comes to editing, particularly when the assembly of multiple inserts is involved. Magnetic disks, designed for computer technology, are well suited to provide convenient access for digital audio. (The spelling 'disk' is standard in computer parlance.)

The floppy disk

As its name suggests, the floppy disk recorder uses a flexible circular disc of material similar to that used for magnetic tape, typically $5\frac{1}{4}$ in or $3\frac{1}{2}$ in in diameter. It is contained in a rigid shell, with sliding covers over the access holes for the read/write head to prevent contamination. For single-sided operation the head operates on the lower side of the disc, which is pressed against the head by a spring-loaded pressure-pad. For double-sided use the pressure-pad is replaced by a spring-loaded second head. The storage capacity of floppy disks is limited to a few megabytes (high-quality digital audio requires one megabyte for about 8 seconds); they are also comparatively slow, which tends to restrict their application for audio to the control of computer-based editing and storage systems, mixing consoles etc.

The Winchester disk

Unlike the floppy disk, which has the advantage of being physically removable from the machine and stored, the Winchester or 'hard' disk normally remains in the computer. It does, however, combine considerable storage capacity with very rapid access, making it very suitable for digital audio recording, particularly in the field of editing and random compilation, eventually to be down-loaded on to a removable medium for storage, possibly a digital tape recorder.

Winchester disk drives use hard disks made of aluminium coated with magnetic oxide. Unlike the floppy disk, where the rotational speed is restricted by considerations of wear due to head-to-disk contact, the Winchester disk rotates at several thousand rev/min, with the head riding just above the surface on a cushion of air.

To allow very rapid access to data several heads are mounted on a positioning arm, so that by selecting the nearest read/write head the search movement is minimalized. When used for editing or selecting passages of digital audio the cues can be virtually instantaneous.

Storage capacity

A typical Winchester disk drive has a capacity of about 300 megabytes, so that it can provide up to about 40 minutes of mono audio storage, or 20 minutes of stereo. Limiting the frequency response to 15 000 kHz (the standard for FM stereo broadcasting) would allow about 30 minutes of recording.

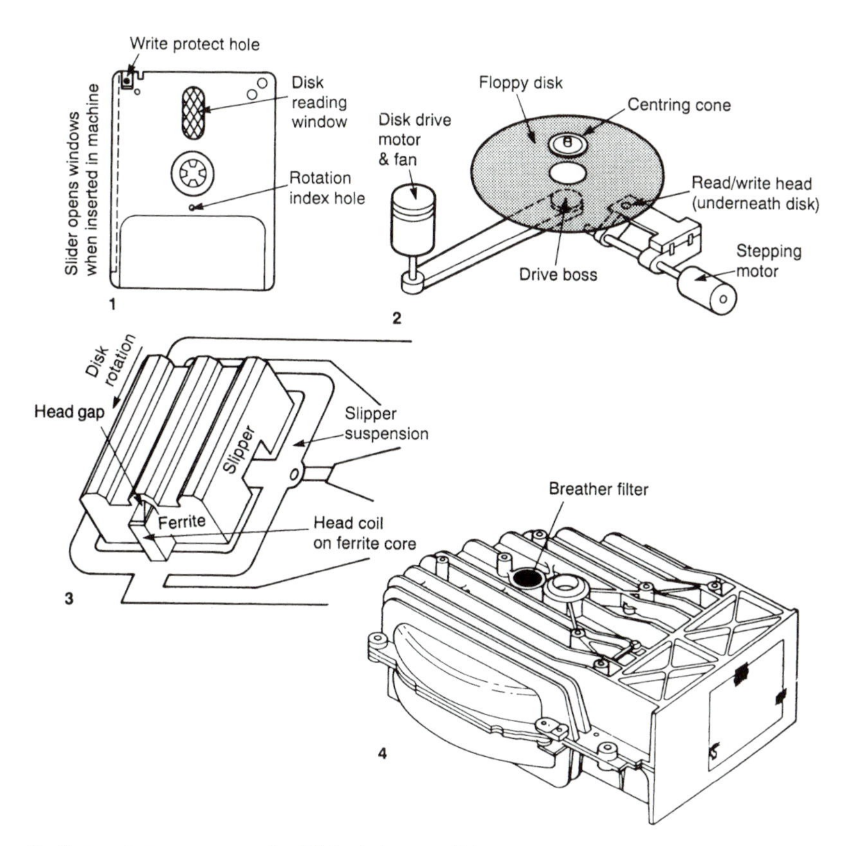

1. General appearance of a 3½ inch floppy disk.
2. General arrangement of floppy disk drive. With a single-side drive the read/write head operates on the lower side of the disk, which is pressed against it by a sprung pressure-pad on the other side. (In a double-sided drive there is another head.) The track is concentric, the heads being moved between tracks by a stepping motor. A small hole in the disk enables a photoelectric sensor to index each revolution of the disk.
3. The Winchester disk rotates at several thousand rev/min, with the heads riding just above the surface on a cushion of air. The separation between the head and the disk is a function of the air movement close to the surface as it is dragged along by the movement of the disk and the aerodynamic shape of the 'slipper' in which the head is mounted. The air cushion increases as the surface of the disk is approached so grooves are provided in the slipper to bleed away some of the air to allow the head, which is lightly sprung, to approach within a few microinches of the disk.

To allow very rapid access several disks are mounted on a common spindle with heads on a rotary positioning arm. It is essential that the heads should never be caused (possibly by vibration or knocks) to touch the working surface of the disk or damage could occur (known as a 'head crash'). For this reason 'parking' areas are provided to which the heads return when not actuated.
4. Due to the very fine tolerances and cleanliness required to maintain the correct disk/head separation, the whole Winchester units are completely enclosed and air sealed, except for a small pressure-equalizing port.

Tapeless Recording and Editing (1)

The availability of high storage capacity computers and disk storage systems and their ability to perform complicated editing and compilation functions entirely in the digital domain has led to a situation where most, if not all, of the audio recording and replay operations can be achieved without the use of tape. One advantage of digital manipulation is the ease with which recordings can be synchronized, e.g. to vision and to each other. Even the individual tracks in a multi-track recording can have their sync adjusted relative to each other, a feature not available in multi-track analog tape recording. Computers can be used to control audio in much the same way as word processors manipulate words, allowing random access to a large number of sources and the ability to shuttle and autolocate cues virtually instantaneously. It is also possible to adjust the length of audio material without altering the pitch or, conversely, to change the pitch without altering the length of the programme material. This may seem to be a contradiction of the fundamental laws of audio, which stipulate that time, wavelength and pitch (frequency) are interrelated. Possibly the best way of understanding this phenomenon is to consider the digital audio signal as a series of samples, the number of which and the space between them being altered without altering the overall shape. It can be achieved by feeding the digital audio into a memory at one clocking rate and reading it out at another.

Operational convenience

An increasing number of devices are being manufactured to meet the needs of the audio recording profession. In general they are designed to be 'user friendly', i.e. they attempt to make complex operations reasonably straightforward and, for the benefit of people who have learned their trade in the pre-digital era, as much like the physical activity of tape handling as possible.

Speed and accuracy

The two main advantages of digital editing devices are speed and accuracy, but in fact one can often be traded off against the other. Digital editing devices can enable the operator to rehearse edit points and hone them to a fine degree which, while it produces very accurate results, can lead to time-wasting changes of mind. A cut and splice edit with a razor blade tends to be final or at least very difficult to remake. A razor blade edit with a 45° angled cut in an analog recording is in fact a cross fade with a timescale of $\frac{1}{60}$ second. This means that the tracks on each half of the tape are cut at a different point, which can cause the stereo image to jump from side to side. Digital editing eliminates this problem and it is even possible to arrange to have different fade times (and amount of overlap) on each side of the cut.

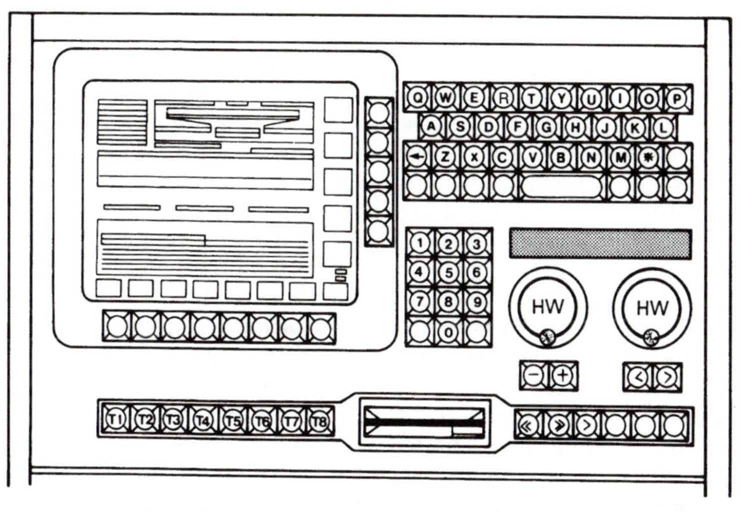

A large number of tapeless recording and editing devices are now available. In general they provide multi-track recording to hard disk, with comprehensive editing and audio manipulation functions. As the hard disks have to remain in the machine it is usually necessary to down-load the finished product to some removable medium such as DAT or compact disc.

Various methods of control are employed by different models, some of which are illustrated here and overleaf.

The Audiofile (above) can perform many functions such as cut and splice editing with variable slope cross-over. It can provide up to eight tracks, programmed from a 'file' of pre-recorded material instantly available in any order of sequence. It can be used for mastering CDs. It generates its own time code, or can be synced to other sound or video sources using the SMPTE/EBU time code.

The controls are arranged around a monitor screen and consist of 'soft' keys the function of which is displayed on the screen adjacent to each key, changing according to the system in use at the time. There is also a typewriter keyboard to type in cue titles and instructions to the computer and a numeric keypad to key in time code information.

For editing, the programme in the area of the edit point is dumped into the memory, access to which is controlled by hand wheels. Turning the hand wheels enables the recording to be rocked backwards and forwards and the edit point trimmed precisely.

The Audiofile features a number of software 'pages' which allow audio to be recorded and edited across eight tracks. This is illustrated graphically on the screen as tracks of an imaginary multi-track recorder. One software package enables the machine to be used as a replacement for cartridge machines, allowing sounds to be arranged in up to eight stacks, selected and manually or automatically cued instantly.

Various alternative methods of manipulating recordings for editing are offered by different manufacturers.

Tapeless Recording and Editing (2)

The Synclavier 'Direct-to-disk'

The Synclavier was developed for the music recording world, with a 76-note piano-style keyboard. Used in conjunction with the 'Direct-to-disk' recording/editing system it is capable of performing most of the functions of a complete multi-track recording studio without use of tape, with the ability to secure rapid acess to material for synchronizing and cueing-in large number of inserts. The synclavier keyboard is velocity and pressure sensitive, the two characteristics being available to cue different effects instantly. Another method of control is a 'mouse' controller which, in association with a vision monitor, enables the operator to select an exact point in an audio waveform which can then be modified, edited, or erased. In this way unwanted clicks and glitches can be removed.

The computer system has available a maximum configuration of 8 gigabytes of on-line storage capacity. Winchester disks provide on-line availability for projects of libraries of frequently-used sounds. Random access memory is used for the recording, editing and playback of short instrumental sounds and sound effects. The Direct-to-disk recorder can be configured in 4, 8 and 16-track units. It records and plays back audio from a network of Winchester disks, giving a maximum recording time of over three hours.

Sonic System

The Sonic System 'Desktop Audio' (from Sonic Solutions) is an editing and signal processing system based around the Apple Mackintosh II computer. The computer is controlled by a keyboard and a 'mouse', in association with a 'Sonic Signal Processing' card. A library of cards is available, and continually updated, containing routines for editing, filtering and mixing etc. The functions in use are graphically displayed on the screen, enabling the operator to control the operation visually with the use of the mouse. The SSP card has the ability to split its four signal processors to handle multiple tasks simultaneously.

The Tablet

The Tablet has an upwards-facing flat screen that is touch sensitive. Its display shows a representation of volume level meters and function buttons. Pressing the picture of a button produces a click (purely to indicate that the action has been accomplished). While the machine is playing there is a representation of the recording moving along a diagrammatic 'tape'. Each edit point is shown as a label which appears at the left, moves along the tape and drops off the right-hand end. A series of pages are available; each provides a different function illustrated by a different graphic display.

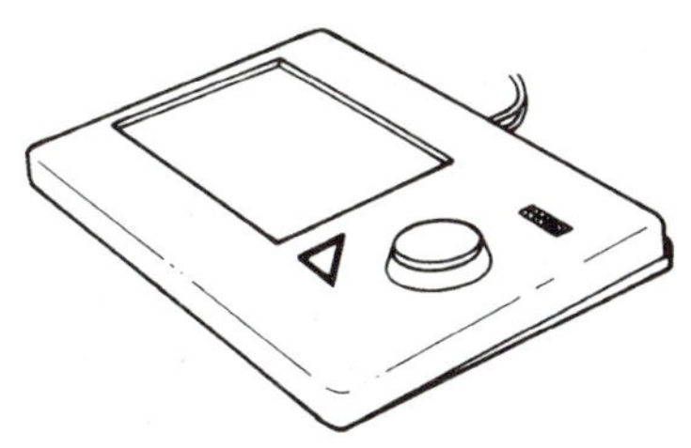

1

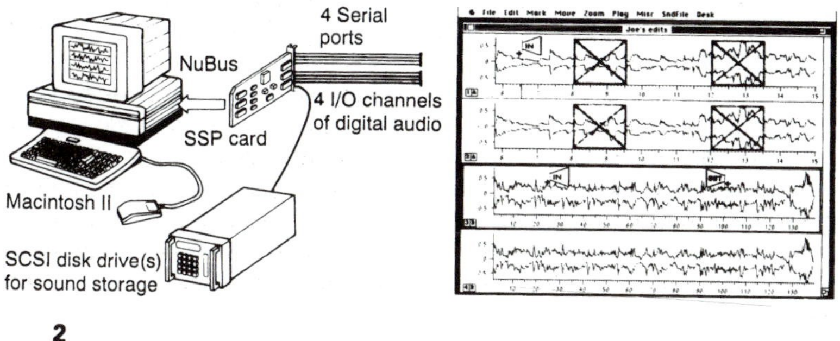

2

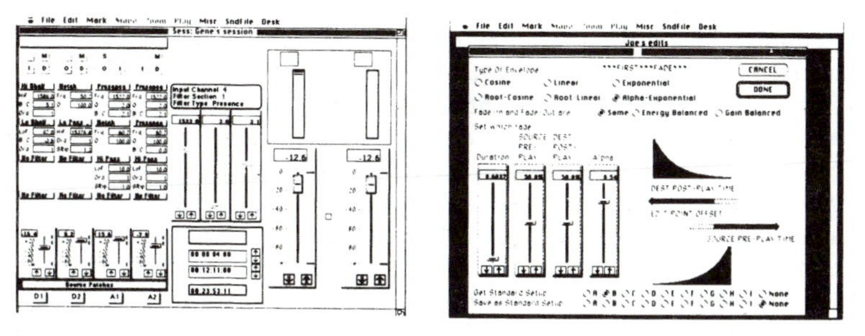

3

There are a large number of tapeless editing systems available for processing and editing sound, and new devices are continually coming onto the market, offering a wide range of facilities and methods of operation.

Illustrated are some examples of different approaches to sound control and editing.

1. The Synclavier.
2. Sonic System showing 'pages' for editing, signal processing and fade control.
3. The Tablet.

Audio recording with no moving parts.

Motionless Recording

The audio signal has two basic coordinate elements: amplitude and time. In the recording systems we have considered so far the time element has been embodied in mechanical movement, whether it be the stylus in a moving groove, tape passing a head or a laser reading a moving disc.

However, it is possible to record and reproduce digital audio signals entirely electronicaly *with no moving parts* using RAM (Random Access Memory). RAM can be considered as a vast number of microscopic capacitors interconnected by a lattice of transistor switches, addressed and controlled by an electronic clocking system. Information can be stored and retrieved (clocked in and out) at the appropriate rate.

NAGRA ARES-C

The NAGRA ARES-C is a solid-state recorder designed principally for broadcast reporters working in the field. It is lightweight and shock-resistant and similar in appearance to the range of NAGRA tape recorders which is very well established in film, television and broadcast circles.

The NAGRA ARES-C has no moving parts or tapes: it uses credit-card-sized, low power, PCMCIA Flash RAM cards for audio storage. The 16-bit, solid-state audio recorder/player uses G.722 data compression to store up to 4 hours 16 minutes (mono) or two 64 Mbyte PCMC1A cards plugged into its double-slot card holder. A 64 Mb card will give a recording capacity of just over an hour in stereo and a 20 Mbyte card approximately 20 minutes. This represents an enormous amount of storage, so that data compression is required, the length of recording being traded off against frequency response. In the G.722 mode the frequency range is restricted to between 30 Hz and 7 kHz (–3 dB). (The MPRG II Compression standard will become available as an option later.) This is likely to be quite adequate for reporting, where only speech is involved, but no doubt if the principle becomes adopted for music recording a different configuration will be devised and cards with even greater addressable capacity will become available. Continuous recording is possible with the ARES-C by loading both card holders, then when one card is full the machine automatically switches to the other and the first card can be replaced.

Copying from one card to the other is possible and full virtual (non-destructive) editing is available, including accurate cut and paste and assembly. A visual representation of a tape appears on the back-lit LCD screen showing edit points, including markers which can be placed at any time during recording without affecting the continuity of the recording. Solid-state recording requires half the battery power of a hard drive of similar capacity, an important factor for outside use.

WORKING ENVIRONMENT OF THE NAGRA ARES-C

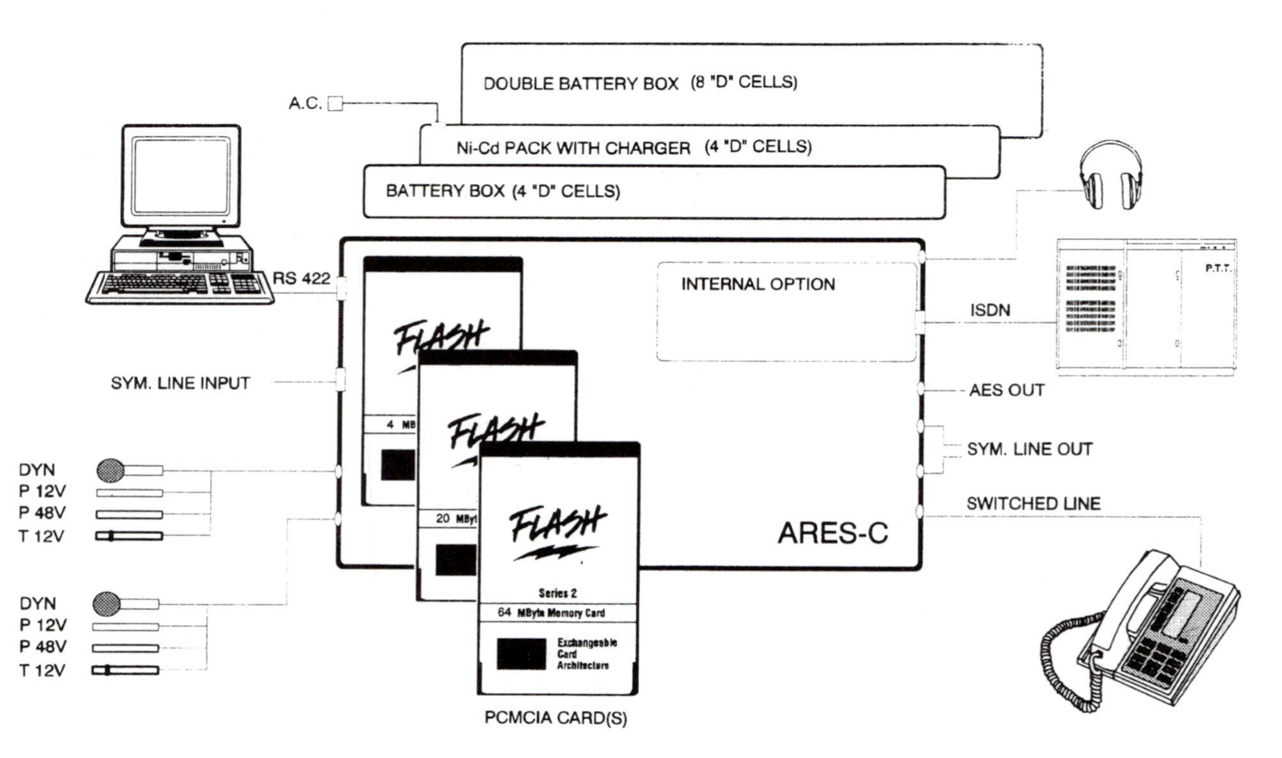

The last, and often the weakest, link in the sound reproduction chain is the loudspeaker. Problems stem from the wide range of wavelengths involved.

Loudspeakers

Moving-coil loudspeakers

The vast majority of present-day loudspeakers are of the moving-coil type. These consist essentially of a diaphragm attached to a small circular (speech) coil which is suspended in a small annular gap between the polepieces of a powerful magnet.

Alternating current applied to the coil produces a varying magnetic flux around it. This flux interacts with the field from the magnet to cause it and the attached diaphragm to move in and out.

The diaphragm can take the form of a dome within the circumference of the coil and/or a cone surrounding it. The dome shape acts as a piston and is suitable for high frequency reproduction. The action of the cone is more complex and varies with frequency. The cone only behaves as a true piston for frequencies with a wavelength well below half its diameter (typically about 400–500 Hz). As the frequency increases the radiating area moves towards the centre, the outer area of the cone acting as a flexible mounting and as a horn-loaded termination. The contour of the cone and the stiffness of its construction, therefore, have a considerable bearing upon the upper frequency response. The low frequency response is largely conditioned by the mass/stiffness of the cone, the flexibility of the outer and inner suspensions and, of course, the baffle in which it is mounted.

Multi-unit loudspeakers

The problem of maintaining loudspeaker diaphragm efficiency throughout the audio frequency range has led to the introduction of multi-unit loudspeakers in which the spectrum is divided into two, three or four frequency bands each handled by a different unit specifically designed for its range. Large cone loudspeakers with free suspensions allowing large cone movement are used for the bass, smaller cone units are used for the mid range, and small flat or dome-shaped diaphragm units for the high frequency. The loudspeaker units are each supplied via a crossover filter network.

Motional feedback loudspeakers

Reasons of economy dictate that most loudspeakers have passive crossovers, i.e. after the power amplifier, but this creates problems of matching. Loudspeaker impedances vary, increasing with frequency. An elegant way of overcoming this problem is to use separate amplifiers for the units in which the feedback circuits of the amplifier are controlled by the voltage developed in a special ceramic pick-up mounted on the cone but not physically connected to anything else. The control emf is produced by the rate of acceleration of the cone acting against the inertia of the pick-up. The output thus compensates for loudspeaker resonances and environmental effects.

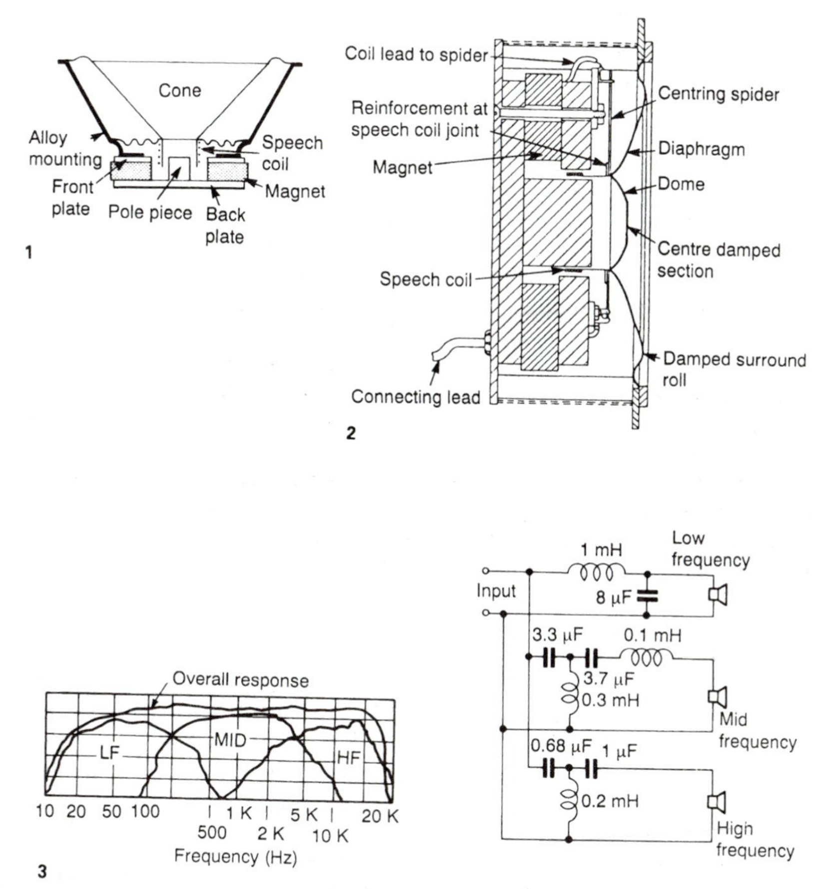

1. Cross section of typical moving-coil loudspeaker.

2. Alternative magnet assembly, suitable for an HF unit. The diaphragm is dome shaped.

3. Typical response curves for a three-unit loudspeaker system.

4. Typical crossover unit.
Various circuits can be used giving various degrees of steepness of cut off. First order filters give 6 dB per octave attenuation, second order 12 dB, and third order 18 dB. Low order filters tend to provide less phase change but produce more overlap between the loudspeaker units.

To obtain bass response from a loudspeaker, it must be surrounded by a baffle or mounted in an enclosure.

Loudspeaker Enclosures

A loudspeaker diaphragm radiates sound waves from the front and the back, the two outputs being 180° out of phase with each other (while the front is compressing the air, the rear is rarefying it). Thus, unless measures are taken to prevent it, the pressure waves will rush around from the front to the back to equalize the pressure and cancel out the sound radiation. The lower the frequency, the more time there will be for the pressure equalizing action, so the less efficient will be the loudspeaker.

Baffles

Ideally each side of the loudspeaker should be working in a totally separate air space, e.g. in the wall between two rooms, but this is not often a practical arrangement and so a baffle is used around the loudspeaker cone. To be effective, the smallest dimension of the baffle must be at least half that of the longest wavelength it is desired to reproduce, e.g. 3.3 m at 50 Hz. This is obviously impossibly large for most situations and it is much more convenient to fold the baffle into the form of a box.

Infinite baffles

If a loudspeaker is placed in a completely enclosed, air-sealed box the energy at the rear of the cone will be taken up in overcoming the stiffness of the air in the box. This will have the effect of increasing considerably the system resonance of the loudspeaker, and units designed for this purpose have a very low free-air resonance. The efficiency can be improved by almost filling the box with absorbent material and by increased magnetic damping in the drive unit. Both these measures will broaden the spectrum of the resonances and make them less pronounced. The absorption of energy at the rear of the loudspeaker has to be paid for in loss of radiating efficiency but this is not of much importance with modern transistor amplifiers.

Vented enclosures

A method of obtaining extended bass response is to provide a vent in the enclosure. This effectively makes the enclosure act as an Helmholtz resonator with a resonance frequency below the cut-off frequency of the unvented enclosure.

Various types of vent are used, such as a small hole through the front of the cabinet or a larger hole connected to a tube to increase its length. Alternatively, an acoustic resistance or a passive unit (a diaphragm with no driving mechanism) can be incorporated in the vent. In this case the effect is to invert the phase of the radiation from the back of the loudspeaker at low frequencies so that it emerges from the box in phase with the front radiation.

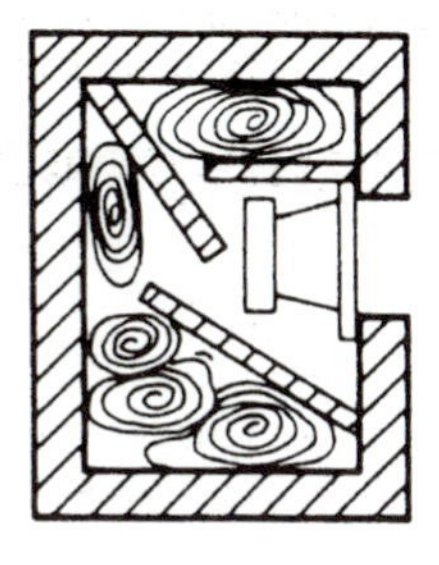

Infinite baffle enclosure containing absorbent material, e.g. fibreglass or rock-wool rolls, and baffles to extend the path length and reduce internal reflections.

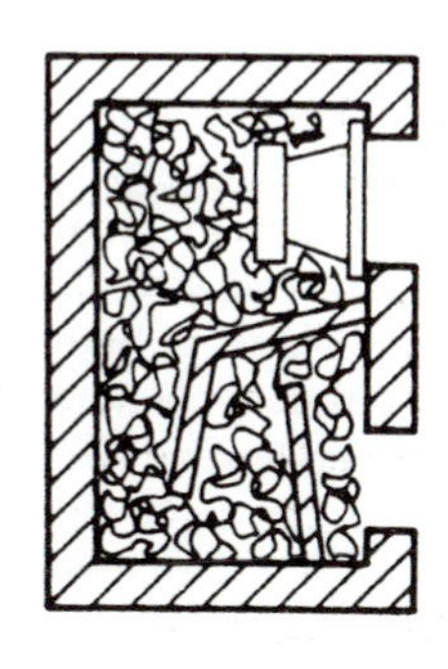

Enclosure with small vent and transmission line loading, i.e. the rear of the loudspeaker is terminated by a tapered duct about one eighth of a wavelength long, lined with absorbent material and terminated with a small vent.

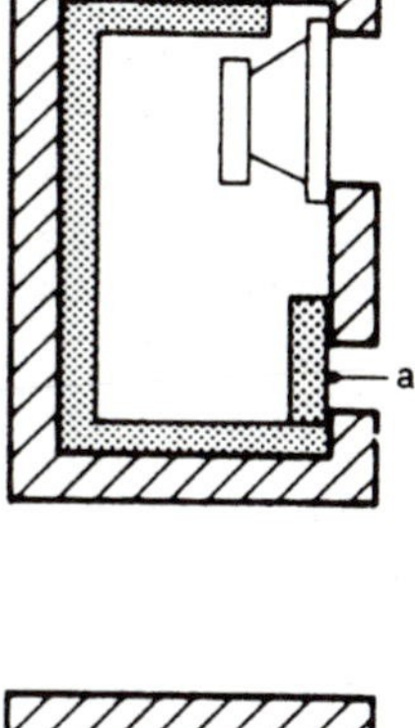

Vented enclosure with acoustic resistance. The inside of the box is lined with absorbent material (*a*).

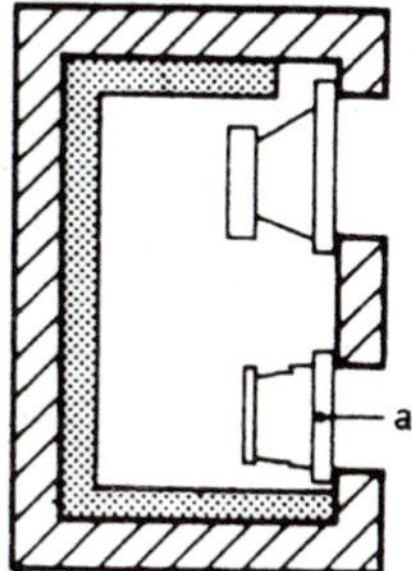

Loudspeaker with a large vent covered by a passive unit (*a*).

Electrostatic loudspeakers can produce excellent sound quality with a rather limited volume output. They must be carefully sited for optimum results.

Electrostatic and Plastic Film Loudspeakers

The electrostatic loudspeaker works on the principle of electrostatic attraction of unlike polarity and repulsion of like polarity. They have been manfuactured with a single fixed backplate and diaphragm but these tend to have a very restricted frequency range.

The double-sided electrostatic loudspeaker

The double-sided push-pull electrostatic loudspeaker is capable of producing very high-quality sound. Basically it consists of a thin conductive-coated diaphragm sandwiched between two rigid perforated backplates.

A balanced, high, polarizing voltage is applied between the diaphragm and the two backplates through a high resistance. The audio signal is applied to the two backplates through a high-ratio step-up transformer. The alternating potential applied by the signal causes the backplates to become alternatively of like and unlike potential with the diaphragm. The diaphragm is then attracted by one plate and repelled by the other alternatively and moves in the space between the fixed plates, which are perforated to allow the resultant sound waves to radiate.

As the diaphragm approaches the fixed backplates, the spacing diminishes and the capacitance increases; this would increase the action, causing a non-linear response. This is the purpose of the high resistance in the polarizing circuit. It is effectively in the charge circuit of the capacity and, by giving it a high value, the time constant can be made longer than the period of the lowest frequency it is required to reproduce. The charge on the plates is prevented from changing at audio frequency and the action is linear.

In view of the high voltage involved it is necessary to provide sufficient stiffness in the diaphragm and limit the force applied to ensure that it does not touch the backplate. As an added precaution the plates can be coated with an insulating plastic film.

In the Quad Acoustical electrostatic loudspeaker there is a separate area in the centre for high frequencies, supplied from a high-pass filter. The full area radiates the bass.

Plastic film loudspeakers

The availability of thick flexible plastic film on which a conductive coating can be printed has made possible the production of some novel forms of loudspeaker which, although operating on the electrostatic principle, use the whole conducting area as a diaphragm in much the same manner as electrostatic loudspeakers.

1. Sectional view of a push-pull electrostatic loudspeaker. The electrodes are made of plastic and only the outer surfaces are metalized.

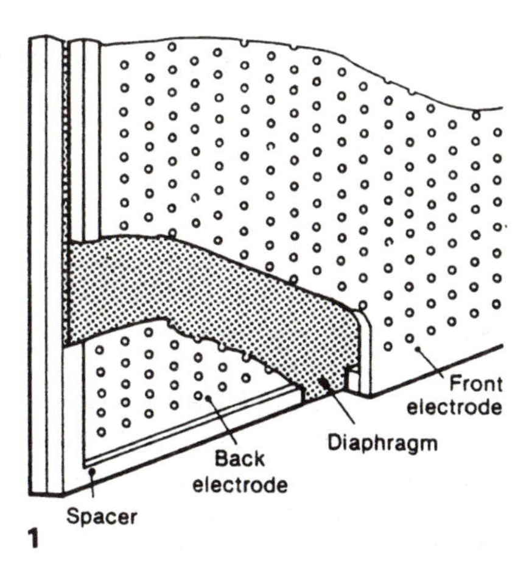

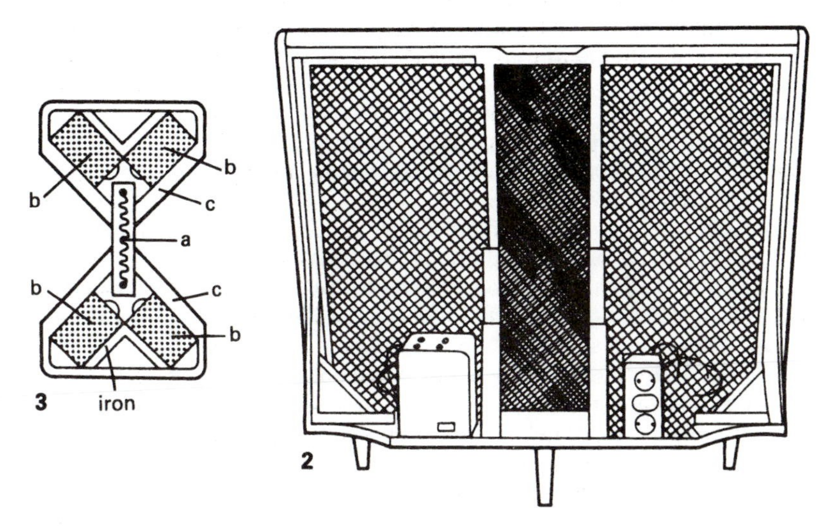

2. Rear view of the Quad Acoustical full-range electrostatic loudspeaker. The polar response is bidirectional at low frequencies, becoming unidirectional at high frequencies.

Correct positioning of the loudspeaker within the room is important. It should not be used close to a wall.

3. Basic construction of a Heil loudspeaker driver. A corrugated conductive film element (*a*) is sandwiched between two powerful magnets (*b*) and polepiece assemblies (*c*). When an audio signal is applied to the element, the corrugations alternatively open and close in a bellows action, modulating the air forced through the gap between the polepieces.

The Dangers of High Sound Levels

No book about sound recording would be complete without mention of the dangers of exposure to excessive loudness levels. Medical evidence has confirmed that long exposure to very loud sound can cause damage to the hearing mechanism, resulting in permanent deafness.

The range of human hearing

Our ears encounter an enormous range of audio power. It could vary from about 0.000 000 001 W for a soft whisper to 10 000 W for a turbo-jet engine (i.e. 30–170 dB relative to 10^{-12} W).

To cope with this range, the human ear has a built-in limiting mechanism. When subjected to a loud sound our ears switch in a sort of attenuator so that we become partially deaf.

If the sound is not too loud or the exposure too long, our hearing recovers soon after the sound has stopped. The louder the sound and the longer the exposure to it, the longer it will take for our hearing to recover. This could take a few hours or days. In the extreme case, where exposure to loud sound has been long and not interspersed with long periods of rest, physical damage to the cochlea will result in permanent deafness.

One of the dangers of exposure to loud sounds is that the resulting deafness can come on gradually and therefore not be noticed until the condition is acute and the situation irreversible. Impairment usually starts with a dip in the 4000 Hz region, which widens and deepens until there is a general loss of intelligibility.

There is a definite relationship between loudness, duration of exposure and the effect on hearing. Although it varies between individuals, the relationship between duration of sound and hearing impairment is generally linear but the effect of loudness rises as loudness increases and above 100 dB it goes up very rapidly indeed.

Whereas formerly exposure to very loud sounds was rare and of short duration (except for those engaged in noisy industrial processes), nowadays with the availability and popularity of high-power amplifiers, particularly in the 'pop' scene, it is possible to listen to sound levels of over 100 dB for long periods of time.

Sound levels in clubs often exceed 120 dB and there is evidence to suggest that many young people have permanent hearing impairment as a result.

Recording engineers engaged in recording pop music, whose clients have already damaged their own hearing and therefore demand higher and higher levels, are particularly at risk. If it is your profession, your livelihood could be at stake.

Sound power and power level scales

Power (W)	*Power level* (dB re 10^{-12} W)	*Source*
25–40 million	195	Saturn rocket
100 000	170	Ram jet
10 000	160	Turbo-jet engine with afterburner
		Turbo-jet engine, 7000lb thrust
1 000	150	4 propeller airliner
100		4 propeller airliner
10	140	75 piece orchestra { peak RMS levels in 0.125 s sec intervals
10	130	Pipe organ { peak RMS levels in 0.125 s sec intervals
1	120	Piano
0.1	110	Blaring radio
0.01	100	Car on motorway
0.001	90	Voice shouting (average long time RMS)
0.000 1	80	
0.000 01	70	Voice, conversational level (average long time RMS)
0.000 001	60	
0.000 000 1	50	
0.000 000 01	40	
0.000 000 001	30	Voice, very soft whisper

Values of sound pressure level in specific frequency bands which indicate a hazard to hearing at stated daily durations for one exposure

Octave band specified as centre frequency (Hz)	*Sound pressure levels at specified durations* (dB)					
	4 hours	*2 hours*	*1 hour*	*30 min*	*15 min*	*7 min*
63	100	103	106	110	116	122
125	94	97	100	104	110	116
250	90	93	96	100	106	112
500	87	90	93	97	103	109
1000	85	88	91	95	101	107
2000	83	86	89	93	99	105
4000	82	84	88	92	98	104
8000	81	84	87	91	97	103

Plugs and Sockets

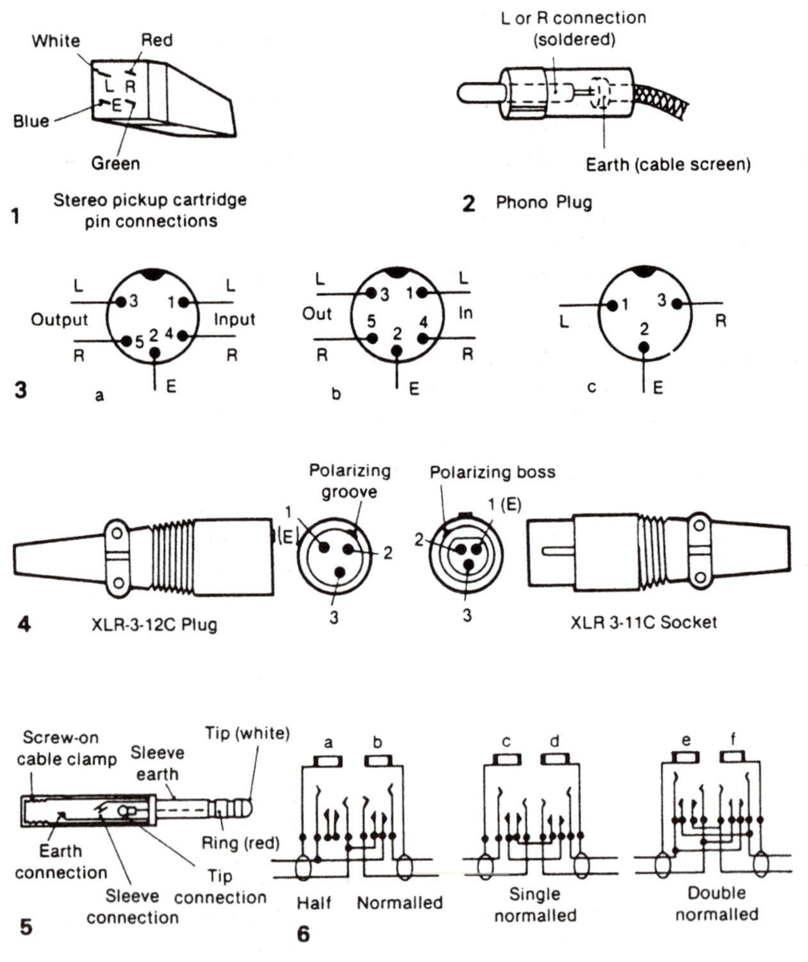

1. Stereo pickup cartridge pin connections.
2. Phono plug. Screened wire should be used for all low-level audio wiring. The screen should be connected to earth at one point only to prevent the formation of an earth loop and the consequent 'hum' problem.
3. Examples of DIN socket connections. Type *a* is the accepted standard for stereo input/output connections. Pin 2 is normally earthed.
4. The XLR3-12C plug and XLR3-11C socket. Pin 1 is normally connected to the cable screen.
5. Jack plug.
6. Jackfield connections. Listening facilities are provided by *a, b, c* and *d* are break jacks. *e* and *f* provide connection in both directions but the through connection is only broken if plugs are inserted in both.

Glossary

Absorption The loss of energy from a sound wave caused by friction with the air or porous material.
AC Alternating current. An electrical current that reverses direction in a periodic manner.
Acoustic treatment The application of absorbent material to the surface of an enclosure to modify the acoustics.
Acoustics The science of sound. The property of an enclosure in relation to sound.
ADC Analog-to-digital conversion.
Aliasing *See* Nyquist limit.
Ambient sound The noise, reverberation or atmospheric sounds that form a background to a sound source.
Amplifier A device by which the strength of a signal can be modified.
Amplitude The peak value of a waveform.
Analog A signal which resembles the original waveform that created it.
Anechoic With reverberation. An anechoic chamber is an enclosure in which there are no audio-wave reflections.
Attack time The time taken for the gain reduction to take effect in a compressor or limiter.
Audio Within the normal range of human hearing, about 20 to 20 000 kHz.
Auto-locate A facility for fast location of a particular part of a recording.
Azimuth The angle between a head gap and the longitudinal axis of the tape. (It should be 90°.)

Balance The achievement of the correct relationship between the various components of a sound source and the acoustic environment to produce the required artistic effect.
Bass The lower frequencies in the audio spectrum.
Beats The periodic variations of amplitude resulting from the combining of two periodic waveforms of similar kind but slightly different frequency.
Bel A scale used to describe the relative magnitude of powers. It is equal to the logarithm to the base 10 of the ratio of the powers. It is used extensively in audio work to describe the level of a signal in relation to a set datum. For audio purposes the bel is too large a unit and the decibel (one tenth of a bel) is used. One dB represents just about the smallest change in audio level discernible to the human ear.
Bias In magnetic recording, the superimposition of a steady magnetic field on the program signal magnetic field to bring the signal excursions to a straight part of the magnetization transfer characteristic, thereby avoiding non-linear distortion and increasing the output.
Bias compensation In phonograph tone arms, a small torque is applied, usually by a spring or a small weight on a string, to compensate for the tendency for the arm to be drawn towards the centre of the record by the drag of the pick-up in the groove.

Bi-directional A microphone with a figure-of-eight directivity pattern.
Binaural hearing The ability to perceive the direction of a sound source, either directly or by electroacoustic means whereby the two ears are supplied with separate transmission channels. Microphones for binaural sound are usually embodied in a dummy head, or separated by a small baffle to simulate the natural condition.
Bit Contraction of 'binary digit', the basic digital unit. Eight bits make a 'byte'.
'Bitstream' conversion Instead of measuring the signal 44 100 times a second (which would provide for 65 536 [2^{10}] possible values of analogue level), converts the 16-bit digital signal into a high-speed 1-bit data stream, extracting the most significant bit and measuring the change of level, up or down, effectively doing 256 calculations each second on each of the 44 100 digital words available, creating a single bitstream with a frequency of 11.3 MHz, equivalent to 256 times oversampling. The intention is mainly to improve the low-level response where errors in significant bits could cause serious distortion. However, the process causes its own degree of noise and distortion but this is spread over the whole 11.3 MHz spectrum and can be shifted from the audio spectrum by 'noise shaping', using the 15 'discarded' bits.
Bulk eraser A device for erasing a complete tape or magnetic disc in a few seconds by submitting it to an AC magnetic field.

Capacitor A component made up of two conducting plates separated by insulation. Capacitors present a decreasing impedance to alternating current as the frequency is increased.
Capacitor microphone Also called condenser or electrostatic microphone. A microphone in which the pressure of sound waves varies the capacitance between the diaphragm and a fixed back plate. It is fed with a polarizing voltage in series with a high resistance across which the output is generated.
Capstan (puck) Drive spindle of a tape machine. Traction is achieved by the pinch wheel pressing the tape against the capstan.
Cardioid microphone A microphone with a heart-shaped directivity pattern.
Cartridge Disc reproducing head, or tape magazine.
Cassette A tape-handling system in which the tape is spooled between two reels contained in a plastic box.
CCIR An international standard of the Comité Consultatif International des Radiocommunications.
Chromium dioxide A type of magnetic tape coating which permits the recording of higher levels of high frequency than regular (ferric oxide) tape, but requires a different value of bias.
Coercivity The ability of a substance (e.g. magnetic tape) to accept and retain magnetism.
Coincident pair An arrangement of microphones for stereophony, in which two microphones are placed with their diaphragms close together, possibly in the same housing, but facing different ways.

Compact disc (CD) A pre-recorded audio record, which is played by scanning with a laser beam. It is very robust and is capable of producing very high quality, with very low inherent noise. Some CDs also provide video.

Companding The application of compression to a signal before a process (e.g. recording); and expansion afterwards to reduce the system noise of the process.

Compressor A variable-gain amplifier, in which the gain, above a set threshold, is controlled by the input signal. It is used for reducing the dynamic range of the signal.

Crossover frequency The frequency at which equal power is delivered to both outputs of a dividing network. In the case of loudspeakers the same power would be applied at this frequency to both the 'woofer' (low-frequency loudspeaker) and the 'tweeter' (high-frequency unit). Above and below this frequency the two units begin to take over, one from the other.

Crystal (piezo-electric) A crystalline substance (usually of rochelle salts, quartz or tourmaline) which has the property of generating small voltages when physically stressed. Used in some microphones and phonograph pick-up heads.

Cycle One complete sequence of a variation which occurs in a periodic manner. One cycle per second is one hertz (Hz).

DAC Digital-to-analog converter.

DASH recorder Digital audio recorder with stationary head. A very high-quality recording system, often used for mastering phonograph records.

dBA A scale, used for measuring noise, that takes into account the unequal sensitivity of our hearing to the audio frequency spectrum.

DBX A noise reduction system which employs pre-emphasized compression before recording and de-emphasized expansion afterwards.

Decibel *See* bel.

De-emphasis A response that decreases as frequency increases. (The opposite of pre-emphasis.)

Digital sound A process for dealing with audio in which the original analog waveform is converted into a series of numerical measurements which can be described by a digital code, usually binary, for which the processing equipment only has to recognize two states: 1 (on) or 0 (off).

Dolby A noise reduction process named after its inventor, Dr Ray Dolby. The Dolby A system is a frequency-selective companding arrangement, used before and after professional recording equipment especially for the production of master tapes. Dolby B and C are processes applied to commercial records on the assumption that complementary equipment will be used for replay. Dolby SR (Spectral Recording) is similar to Dolby C except that three levels of action staggering are used: high, mid and low level. The main principle is to apply the companding action only at the frequencies where it is needed at each instant and to keep the low-level components of the signal boosted at all times.

Dropout Momentary loss of signal due to a fault in a tape coating or dust causing the tape to lift off the head.
Dubbing The process of re-recording from one recording to another. Sometimes used for the addition of sound (e.g. foreign dialogue) to a previously recorded picture.
Dynamic range The ratio between the softest and loudest sound. Or the ratio of the signal variation available between unacceptable noise level and overload. (Expressed in decibels.)

Echo Discrete, separately identifiable repetitions of sound due to reflection of sound waves from hard surfaces. The term is often used, wrongly, to mean reverberation, which is due to multiple overlapping reflections, which blend together to prolong the sound.
EFM Eight-to-fourteen modulation, a technique which converts each 8-bit into a 14-bit symbol, with the purpose of aiding the recording and playback procedure by reducing the required bandwidth, reducing the signal's DC content and adding extra synchronizing information. The procedure is to use 14-bit codewords to represent all possible combinations of the 8-bit code. An 8-bit code represents 256 (i.e. 2^8) possible combinations. A 14-bit code represents 16 384 (2^{14}) different combinations. Out of the 16 384 14-bit codewords only 256 are selected, having combinations which aid processing of the signal. For instance, by choosing codewords which give low numbers of individual bit conversions (i.e. 1 to 0 or 0 to 1) between consecutive bits the bandwidth is reduced. Also by choosing codewords with only limited numbers of consecutive bits with the same logic level the overall DC level is reduced.
Eigentones Standing-wave resonances in an enclosure when the reflective path length coincides with the wavelength of a sound.
Equalization (EQ) The process of modifying the amplitude/frequency response to correct for amplitude distortion or to minimize noise. The term is also used for the deliberate distortion, or shaping, of a sound signal for artistic effect.
Erasure The process of removing previously recorded signals from magnetic materials prior to recording.

Flutter A periodic pitch variation with a repetition rate greater than 10 times per second. A slower repetition is called 'wow'.
Frequency The rate of repetition of a periodic function, measured in *hertz* (Hz).
Frequency modulation A method of modulating a carrier by varying its frequency as opposed to its amplitude.

Gain Amplification of a component. Usually expressed in decibels.
Guardband Spacing between tracks on a recorded tape.

Haas effect The effect that determines the apparent direction of a source of sound. When the same sound is reproduced simultaneously from two or more places it will appear to come entirely from the nearest one, unless

another source is very much louder. The relationship between sound volume, time delay and directionality is known as 'Haas effect'.
Harmonic A sinusoidal oscillation having a frequency which is an integral multiple of the fundamental frequency. It is the harmonics (or in musical terms upper partials) that shape the waveforms and make it possible to distinguish between various instruments, even when they are playing the same note.
Harmonic distortion The production of spurious harmonics due to irregularity in signal processing distorting the original waveform.
Hertz (Hz) Unit of frequency. One hertz equals one cycle of repetition per second.
Hypercardioid Directional response (polar diagram) between cardioid and figure-of-eight. For example a microphone with hypercardioid response would have a narrower than cardioid acceptance angle in front, with a small lobe at the back.

Impedance The restriction of current due to the total resistance and reactance of a circuit.
Intensity of sound A measure of the power of a sound source; usually measured in decibels relative to the threshold of hearing at 1 kHz.
Intermodulation distortion A form of distortion caused by the mixing of component frequencies, in which one frequency modulates another, causing the production of spurious tones and a general roughness of the sound.

Laservision A digital video recording on disc. The system of playback is similar to CDs, using a laser beam. CD video players connected to television and stereo systems can play audio CDs or video discs to provide vision and stereo sound. Three types of CD video discs are available: (1) 12 cm singles, single-sided, containing up to six minutes of vision with stereo sound, plus up to a further 20 minutes of sound only, which can be played on a normal audio CD player; (2) 20 cm double-sided discs with up to 20 minutes of video and sound per side; (3) 30 cm long-play double-sided discs with up to 60 minutes of vision and sound on each side.
Level The intensity of steady tone, usually of 1 kHz, used for test purposes and for lining-up equipment. Line-up (zero) level is usually quoted as a power of *i* mW in a resistance of 600 ohms.
Limiter A device for preventing the volume of a signal from rising above a pre-set value, to allow a higher average level to be accommodated without overload.
Loudness The subjective auditory impression of the intensity of a sound.

MADI Multi-channel audio digital interface. Carries a number of channels of audio over a single coaxial cable or optical fibre.
Magnetic flux The lines of force emanating from a magnet.
Masking The process by which the ability to hear sound of a particular frequency is reduced by the presence of louder sounds of similar frequency.
Mega Prefix signifying one million (M).

Micro Prefix signifying one millionth part (μ).
Microphone A transducer for converting sound waves into electrical signals.
MIDI Musical instrument digital interface. A standard interface protocol, used to link instruments and computer systems, e.g. electronic keyboards, drum machines, and audio processing devices.
Modulation The control of one waveform by another.
Monaural, monophonic, or mono A single-channel sound system.
Moving coil A transducer system based upon the movement of a coil of wire in a magnetic field.
Mumetal An iron alloy used for tape heads and general magnetic screening.

NARTB Standards for recording and reproduction of the National Association of Radio and Television Broadcasters.
Nyquist limit In converting an audio signal between analog and digital format the sampling frequency must be at least twice the highest frequency it is intended to convert. This is the Nyquist limit. If it is not exceeded a form of heterodyning, called *aliasing*, will occur, caused by beating between the input and the sampling frequency. In practice the sampling rate has to be at least 2.2 times the highest audio frequency to allow for filtering out the out-of-band components of the audio signal.

Obstacle effect Sound waves are reflected or absorbed only by obstacles whose dimensions are larger than their wavelength. Longer waves merely curve around them, especially if streamlined in shape.
Ohm Unit of resistance or impedance.
Ohm's Law General law describing the flow of direct current in an electrical circuit. It states that the magnitude of the current (in amperes) is proportional to the potential difference (in volts) divided by the resistance (in ohms).
Omnidirectional Acting equally in all directions.
Oversampling When converting between digital and analog it is necessary to use a sampling rate at least twice the highest frequency it is intended to convert (see Nyquist Limit). Oversampling is a method of increasing the effective sampling rate by interpolation.

Pan pot Panoramic potentiometer. A control used in stereophony for adjusting the apparent lateral direction of a sound source.
Parametric equalizer A frequency-shaping device, which can boost or reduce a signal at a particular frequency. The position on the scale and the width of the peak or trough in the response can be varied.
Peak programme meter A meter for measuring programme volume. it has a rapid rise time and slow recovery, so that it is easy to read peak values.
Phase The position in the cycle that a waveform has reached at any given instant. Waves are said to be in phase when their position in the cycle coincides.

Phon A measure of the loudness of a sound that takes into account the unequal sensitivity of our ears to sounds of different frequencies. The decibel scale of aural sensitivity (equal loudness/frequency) is the same as the phon scale at 1000 Hz, but differs at other frequencies.
Pick-up (or pick-up cartridge) A transducer for converting the contours of a record groove into electrical signals.
Pitch Subjective auditory sensation of sound related to a scale mainly of frequency, but also influenced by such factors as intensity and harmonic structure.
Playback curve Standard equalization characteristic used for playback of tape or disc recordings.
Pre-emphasis Boosting the high-frequency component of a signal before recording, or (FM) transmission, so as to reduce the high-frequency system noise when it is subsequently de-emphasized on reproduction.
Print through The transfer of a recording on to an adjacent layer when the tape is wound on a spool. It can sound like echoes of the louder parts of a recording occurring during preceding or following quieter passages.
Pulse code modulation (PCM) A system for converting the audio signal into a series of pulses. The analogue waveform is sampled at a fixed rate, the amplitude of each sample being represented by a binary word of fixed length (typically 16 bits).

Quadraphony An audio system employing four channels in one medium, intended to be played on four loudspeakers arranged around the listener to provide an all-round effect.

Remanence The ability of a material to retain residual magnetism when the magnetizing force has been removed.
Reverberation The sustaining effect of multiple sound wave reflections within an enclosed area, with reflecting surfaces. It is quantified in the time in seconds that a sound takes to die away to one millionth (–60 dB) of its original power after it has been cut off abruptly.
Rumble A low-frequency vibration in phonographs, usually caused by transmission of vibration from the turntable motor to the record platter and on to the pick-up.

Sine wave The pattern executed by a simple cyclic variation, such as an alternating current. In audio terms it is a pure tone.
Stereophony A two-channel audio system which can provide the impression of spatial distribution of sound sources in the horizontal plane.
Stylus The point of contact that tracks the groove of a phonograph record. Usually tipped with sapphire or diamond.

Tape transfer characteristic The relationship between the applied magnetizing force (H) and the induced magnetic flux (B) in a recording tape. When the magnetizing force alternates in direction the transfer characteristic follows a hysteresis loop.

Time constant A method of stipulating the shape of a frequency-correction curve by the charge/discharge time, quoted in microseconds, of a circuit containing resistance and capacitance.
Tone arm (tracking arm) The arm that supports the pick-up in a phonograph.
Transducer A device for converting one form of energy into another, e.g. alternating electrical signals into mechanism movement, or vice versa.
Tweeter The high-frequency element in a multi-unit loudspeaker system.

VCA Voltage controlled amplifier. The gain of a VCA is controlled by a small DC voltage applied by a fader in a sound control console, thereby reducing the risk of noise caused by poor contact when programme is routed through faders.
VU meter A meter for indicating programme volume.

Waveform The shape of a graph depicting the successive values of a varying quantity.
Wavelength The distance between corresponding parts of a waveform.
White noise A full-spectrum signal having the same energy level over the whole audio frequency range.
Winchester disk A magnetic disk with very high storage capacity, developed originally for computers but also capable of high-quality digital audio recording. ('Disk' is the accepted spelling when associated with computers.)
Woofer The low-frequency element in a multi-unit loudspeaker system.
Wow A slow variation of pitch, normally most noticeable on sustained low notes, usually due to eccentricity in tape recorder or turntable drive system or enlarged centre holes in discs.

Zero level The level used for lining-up equipment using standard tone. it corresponds to a power of 1 mW in 600 ohms. It is indicated by the figure 4 on a peak programme meter.